Intervertebral Disc Regeneration

Intervertebral Disc Regeneration

Editor

Benjamin Gantenbein

MDPI • Basel • Beijing • Wuhan • Barcelona • Belgrade • Manchester • Tokyo • Cluj • Tianjin

Editor
Benjamin Gantenbein
Medical Faculty
Switzerland

Editorial Office
MDPI
St. Alban-Anlage 66
4052 Basel, Switzerland

This is a reprint of articles from the Special Issue published online in the open access journal *Applied Sciences* (ISSN 2076-3417) (available at: https://www.mdpi.com/journal/applsci/special_issues/Intervertebral_Disc_Regeneration).

For citation purposes, cite each article independently as indicated on the article page online and as indicated below:

LastName, A.A.; LastName, B.B.; LastName, C.C. Article Title. *Journal Name* **Year**, *Volume Number*, Page Range.

ISBN 978-3-0365-2754-3 (Hbk)
ISBN 978-3-0365-2755-0 (PDF)

Cover image courtesy of Benjamin Gantenbein.

Contents

About the Editor

Benjamin Gantenbein, Prof. Dr., Phil.nat., is Associate Professor of the Medical Faculty and Group Head of the Bone and Joint Program and the bio lab of Tissue Engineering for the Orthopaedics and Mechanobiology (TOM) group at the Department for BioMedical Research (DBMR) and the Department of Orthopaedic Surgery and Traumatology, Inselspital, at the Medical Faculty of the University of Bern. His research focuses on intervertebral disc repair using biomaterials and scaffolds, mesenchymal stem cells or a combination thereof. He began his career at the University of Bern in evolutionary biology/phylogenetics. He completed his Master of Science degree and PhD at the Computational and Molecular Population Genetics (CMPG) laboratory at UBERN. He was then awarded two fellowships (SNF young scientists and a Marie Curie IHP substitute) to focus on animal molecular evolutionary rates at the Institute of Evolutionary Biology, University of Edinburgh. Finally, he moved to the Genetics and Molecular Ecology Laboratory at Cambridge University, focusing on recombination in animal mitochondria. Before the current assignment, Prof. Gantenbein entered into the field of intervertebral disc (IVD) research at the AO Research Institute in Davos, where he acquired experience in modern concepts of regenerative medicine. He is active in tissue engineering and stem cell research. He is further researching bioreactor design using a more comprehensive range of bone and joint tissues. Since 2009, he has been teaching "Tissue Engineering" as part of the Biomedical Engineering Masters course program at the Medical Faculty of the University of Bern. He also teaches mechanobiology at the University of Basel and Fribourg as a guest lecturer of their Biomedical Engineering Programs. His current research focuses on the use of biomaterials such as silk for the regeneration of IVD and on stem cells/ progenitor cells (https://p3.snf.ch/project-192674, accessed 11 Dec 2021). One of his significant achievements in mechanobiology is developing a unique bioreactor device that allows two-degree-of-freedom loading for IVD organ explants under sterile conditions. The entire research around a clinically relevant organ culture involving torsion and compression of the IVD has attracted worldwide attention in the scientific community. As for more recent applied scientific work, his group won a series of notable awards in the field of cells and materials for developing clinically relevant degeneration models for the intervertebral disc (among which includes the prestigious Hansjörg Wyss Startup grant). Furthermore, tight collaborations with spine surgeons at the Insel University hospital led to further promising results using the substance L51P, a bone morphogenic protein (BMP)-2 analogue, and its potential for improving spinal fusion in an in vivo animal model. Further efforts involve the development of induced pluripotent stem cells (iPSCs) using primary cells positive for an angiogenic marker (Tie2/TEK receptor tyrosine kinase; Tie2 or syn angiopoietin-1 receptor) in a collaborative consortium project called "iPSpine" (https://cordis.europa.eu/project/ id/825925, https://ipspine.eu, both accessed 11 Dec 2021), which is coordinated by Prof. Dr Marianna Tryfonidou RMCU, Utrecht University. Finally, a third key topic was recently initiated in Nov 2020, which involves artificial intelligence, statistical shape modeling and finite element modelling and organ culture models for IVD regeneration: The 4M € funded "Disc4All" project, which aims to tackle the issues through the collaborative expertise of clinicians, computational physicists and biologists, geneticists, computer scientists, cell and molecular biologists, microbiologists, bioinformaticians, and industrial partners (https://cordis.europa.eu/project/ id/955735; https://www.upf.edu/web/disc4all; both accessed 11 Dec 2021). It provides interdisciplinary training in data curation and integration, experimental and theoretical/computational modelling, computer algorithm development, tool generation, and model and simulation platforms to transparently integrate primary data for enhanced clinical interpretations through models and simulations. The consortium is led by Dr. PhD Jérôme Noailly from the Universitat Pompeu Fabra (UPF) in Barcelona, Spain.

 applied sciences

MDPI

Editorial

New Frontiers towards Regeneration of the Intervertebral Disc: On Progenitor Cells, Growth Factors and Biomaterials

Benjamin Gantenbein [1,2]

[1] Tissue Engineering for Orthopaedics & Mechanobiology (TOM), Bone & Joint Program, Department for BioMedical Research (DBMR), Faculty of Medicine, University of Bern, CH-3008 Bern, Switzerland; benjamin.gantenbein@dbmr.unibe.ch; Tel.: +41-31-632-88-15

[2] Department of Orthopaedic Surgery and Traumatology, Inselspital, Bern University Hospital, University of Bern, CH-3010 Bern, Switzerland

Abstract: This Special Issue on intervertebral disc (IVD) regeneration focuses on novel advances in understanding the cell sources and culture conditions of various cell types, i.e., progenitor and IVD cells. The issue consists of seven articles that provide a comprehensive overview of recently applied research insights: (1) into how IVD herniation can be provoked in a controlled in vitro biomechanical testing set-up, (2) how cells can be used for IVD repair, (3) the physiological conditions of IVD cells and (4) how hyaluronic acid could be used for IVD repair, and (5) how nucleus pulposus progenitor cells (NPPCs) and mesenchymal stromal cells (MSCs) shall be cultured and expanded towards a possible cell therapy.

Keywords: biomaterials; tissue engineering; progenitor cells; cell therapy; biomechanics; lumbar disc herniation; GDF5; GDF6; TGFβ; nucleus pulposus progenitor cells (NPPC); losartanon

Citation: Gantenbein, B. New Frontiers towards Regeneration of the Intervertebral Disc: On Progenitor Cells, Growth Factors and Biomaterials. *Appl. Sci.* **2021**, *11*, 11913. https://doi.org/10.3390/app112411913

Received: 8 December 2021
Accepted: 13 December 2021
Published: 15 December 2021

Publisher's Note: MDPI stays neutral with regard to jurisdictional claims in published maps and institutional affiliations.

1. Introduction

Current approaches to regenerate the entire IVD or parts of IVDs are not very successful. Lower back pain (LBP) is a dominant medical problem that affects billions of people worldwide, yet no cure is in close sight. Orthopedic surgery often removes the affected motional segment by so-called "discectomy" and tries to fuse the two adjacent segments to achieve stability, possibly removing the LBP. However, tissue-engineered solutions have been developed to target a biological and, hopefully, more physiological solution to the problem. This Special Issue provides a cross-disciplinary section of "applied science" in this field of biomedical research ranging from biomechanics [1], biomaterials such as hyaluronic acid (HA) [2], and cell physiology [3] to in vitro cell culture studies on how cell phenotypes should be directed towards more IVD-like cells [4–6]. Traditionally, three tissue types are distinguished of the IVD, i.e., the nucleus pulposus (NP), the annulus fibrosus (AF), and the cartilaginous endplates (CEP) [7]. Thus, the articles partially focus on the generation of in vitro NP cells and how to provide a micro-environment for these [2]. In addition, an innovative approach is presented regarding how Angiotensin II Type 1 Receptor Antagonist "Losartanon" could be used to partially reverse inflammation caused by tumor necrosis factor alpha (TNFα) [4].

2. Highlights on the Studies Published in the Present Special Issue: Intervertebral Disc Regeneration

The first highlight to mention is the valuable contribution of the group of Prof. Hans-Joachim Wilke and his team on how the process of lumbar disc herniation can be provoked using in vitro biomechanical testing protocols [1]. Wilke et al. (1999) [8] provided essential knowledge on the in vivo physiological loading patterns in human IVDs following the first observations by Nachemson [9] in 1966. In this Special Issue for the first time, a successful loading regime is presented that can reproduce the herniation of NP material from the inner part of the IVD as in the clinical situation [1].

Furthermore, two excellent reviews are presented in this Special Issue: one is on the usage of HA for the application of IVD regeneration by the group of Prof. Abhay Pandit [2]. This group is extremely knowledgeable on various aspects of biomaterials for multiple applications in musculoskeletal regeneration and provides an excellent review of the studies involving HA for IVD repair. The second review focuses on progress using many allogeneic progenitor cells for cell therapy in ongoing clinical trials by Prof. Inbo Han and his team [10]. Furthermore, two studies were published by Prof. Daisuke Sakai and his team around progenitor cells [5,6]. One manuscript by Sako et al. (2021) [5] reports an exciting approach how NPPCs can be better kept in culture using specialized spheroid culture systems. The second study by the same laboratory team then reports on in vitro screening for the most suitable combination of cytokines to differentiate MSCs towards NP-like cells [6]. This discussion on how to obtain NP-like cells from progenitors, of course, already began much earlier and can be read in more detail in many other articles presented by other research groups, e.g., [11–13]. It was exciting to learn that a combination of growth and differentiation factor 5 (GDF5) and transforming growth factor beta 1 (TGFβ1) was the most promising combination towards an NP-like phenotype.

Funding: The author wishes to acknowledge his current funding sources: This research was enabled by the Swiss National Science Foundation project #310030E_192674/1 (https://p3.snf.ch/project-192674, accessed on 1 December 2021). Further funds were received from the iPSpine H2020 project under the grant agreement #825925 (https://cordis.europa.eu/project/id/825925, assessed on 1 December 2021) and by funds from the Marie Skłodowska Curie International Training Network (ITN) "disc4all" (https://cordis.europa.eu/project/id/955735, accessed on 1 December 2021). Hans Joachim Wilke acknowledges his funding by the German Spine Foundation and partially by the German Research Foundation (WI 1352/14-3).

Institutional Review Board Statement: Not applicable.

Informed Consent Statement: Not applicable.

Data Availability Statement: Not applicable.

Acknowledgments: This publication was only possible with the valuable contributions from the authors, reviewers, and the editorial team of *Applied Sciences*, with special thanks to Frederic Yuan, who was very motivating and supportive.

Conflicts of Interest: The author is a guest editor of the present Special Issue and has no other conflict of interest with any of the published contents of the Special Issue articles.

References

1. Zengerle, L.; Debout, E.; Kluger, B.; Zöllner, L.; Wilke, H.-J. In Vitro Model for Lumbar Disc Herniation to Investigate Regenerative Tissue Repair Approaches. *Appl. Sci.* **2021**, *11*, 2847. [CrossRef]
2. Kazezian, Z.; Joyce, K.; Pandit, A. The Role of Hyaluronic Acid in Intervertebral Disc Regeneration. *Appl. Sci.* **2020**, *10*, 6257. [CrossRef]
3. Borrelli, C.; Buckley, C.T. Synergistic Effects of Acidic pH and Pro-Inflammatory Cytokines IL-1β and TNF-α for Cell-Based Intervertebral Disc Regeneration. *Appl. Sci.* **2020**, *10*, 9009. [CrossRef]
4. Saravi, B.; Li, Z.; Pfannkuche, J.; Wystrach, L.; Häckel, S.; Christoph, C.; Grad, S.; Alini, M.; Richards, R.G.; Lang, C.; et al. Angiotensin II Type 1 Receptor Antagonist Losartan Inhibits TNF-α-Induced Inflammation and Degeneration Processes in Human Nucleus Pulposus Cells. *Appl. Sci.* **2021**, *11*, 417. [CrossRef]
5. Sako, K.; Sakai, D.; Nakamura, Y.; Matsushita, E.; Schol, J.; Warita, T.; Horikita, N.; Sato, M.; Watanabe, M. Optimization of Spheroid Colony Culture and Cryopreservation of Nucleus Pulposus Cells for the Development of Intervertebral Disc Regenerative Therapeutics. *Appl. Sci.* **2021**, *11*, 3309. [CrossRef]
6. Morita, K.; Schol, J.; Volleman, T.N.E.; Sakai, D.; Sato, M.; Watanabe, M. Screening for Growth-Factor Combinations Enabling Synergistic Differentiation of Human MSC to Nucleus Pulposus Cell-Like Cells. *Appl. Sci.* **2021**, *11*, 3673. [CrossRef]
7. Cassidy, J.J.; Hiltner, A.; Baer, E. Hierarchical structure of the intervertebral disc. *Connect. Tissue Res.* **1989**, *23*, 75–88. [CrossRef] [PubMed]
8. Wilke, H.J.; Neef, P.; Caimi, M.; Hoogland, T.; Claes, L.E. New in vivo measurements of pressures in the intervertebral disc in daily life. *Spine* **1999**, *24*, 755–762. [CrossRef] [PubMed]

9. Nachemson, A. The load on lumbar disks in different positions of the body. *Clin Orthop. Relat. Res.* **1966**, *45*, 107–122. [CrossRef] [PubMed]
10. Lee, C.K.; Heo, D.H.; Chung, H.; Roh, E.J.; Darai, A.; Kyung, J.W.; Choi, H.; Kwon, S.Y.; Bhujel, B.; Han, I. Advances in Tissue Engineering for Disc Repair. *Appl. Sci.* **2021**, *11*, 1919. [CrossRef]
11. Colombier, P.; Clouet, J.; Boyer, C.; Ruel, M.; Bonin, G.; Lesoeur, J.; Moreau, A.; Fellah, B.H.; Weiss, P.; Lescaudron, L.; et al. TGF-β1 and GDF5 Act Synergistically to Drive the Differentiation of Human Adipose Stromal Cells toward Nucleus Pulposus-like Cells. *Stem Cells* **2016**, *34*, 653–667. [CrossRef] [PubMed]
12. Stoyanov, J.V.; Gantenbein-Ritter, B.; Bertolo, A.; Aebli, N.; Baur, M.; Alini, M.; Grad, S. Role of hypoxia and growth and differentiation factor-5 on differentiation of human mesenchymal stem cells towards intervertebral nucleus pulposus-like cells. *Eur. Cell. Mater.* **2011**, *21*, 533–547. [CrossRef] [PubMed]
13. Clarke, L.E.; McConnell, J.C.; Sherratt, M.J.; Derby, B.; Richardson, S.M.; Hoyland, J.A. Growth differentiation factor 6 and transforming growth factor-beta differentially mediate mesenchymal stem cell differentiation, composition and micromechanical properties of nucleus pulposus constructs. *Arthritis Res.* **2014**, *16*, R67. [CrossRef] [PubMed]

MDPI

Review

Advances in Tissue Engineering for Disc Repair

Chang Kyu Lee [1,†], Dong Hwa Heo [2,†], Hungtae Chung [2], Eun Ji Roh [3], Anjani Darai [3], Jae Won Kyung [3], Hyemin Choi [3], Su Yeon Kwon [3], Basanta Bhujel [3] and Inbo Han [3,*]

1 Department of Neurosurgery, Keimyung University School of Medicine, Dongsan Medical Center, Daegu 42601, Korea; faslcklee@gmail.com
2 Department of Neurosurgery, Seoul Bumin Hospital, Seoul 07590, Korea; spinesurgery@naver.com (D.H.H.); bregma@bumin.co.kr (H.C.)
3 Department of Neurosurgery, CHA University School of Medicine, CHA Bungdang Medical Center, Seongnam-si 13497, Korea; morolro@naver.com (E.J.R.); anjanianji09@gmail.com (A.D.); kyungjaewon88@gmail.com (J.W.K.); littlechoi88@gmail.com (H.C.); syunkwon@naver.com (S.Y.K.); basantabhujel86@gmail.com (B.B.)
* Correspondence: hanib@cha.ac.kr; Tel.: +82-31-780-1924
† These authors contributed equally to this work.

Abstract: Intervertebral disc (IVD) degeneration is a leading cause of chronic low back pain (LBP) that results in serious disability and significant economic burden. IVD degeneration alters the disc structure and spine biomechanics, resulting in subsequent structural changes throughout the spine. Currently, treatments of chronic LBP due to IVD degeneration include conservative treatments, such as pain medication and physiotherapy, and surgical treatments, such as removal of herniated disc without or with spinal fusion. However, none of these treatments can completely restore a degenerated disc and its function. Thus, although the exact pathogenesis of disc degeneration remains unclear, there are studies examining the effectiveness of biological approaches, such as growth factor injection, gene therapy, and cell transplantation, in promoting IVD regeneration. Furthermore, tissue engineering using a combination of cell transplantation and biomaterials has emerged as a promising new approach for repair or restoration of degenerated discs. The main purpose of this review was to provide an overview of the current status of tissue engineering applications for IVD regenerative therapy by performing literature searches using PubMed. Significant advances in tissue engineering have opened the door to a new generation of regenerative therapies for the treatment of chronic discogenic LBP.

Keywords: intervertebral disc (IVD); regeneration; stem cell; biomaterials; scaffolds; tissue engineering

Citation: Lee, C.K.; Heo, D.H.; Chung, H.; Roh, E.J.; Darai, A.; Kyung, J.W.; Choi, H.; Kwon, S.Y.; Bhujel, B.; Han, I. Advances in Tissue Engineering for Disc Repair. *Appl. Sci.* **2021**, *11*, 1919. https://doi.org/10.3390/app11041919

Academic Editor: Benjamin Gantenbein

Received: 1 February 2021
Accepted: 18 February 2021
Published: 22 February 2021

1. Introduction

Intervertebral disc (IVD) degeneration (IVDD) causes chronic low back pain (LBP), including discogenic back pain, significant health problems, and socioeconomic burden [1]. IVDD-induced discogenic LBP accounts for more than 40% of all LBP cases [2] and is considered as one of the top global causes of disability-adjusted life years [3].

IVDD is a process leading to loss of proteoglycans (PGs), destruction of the extracellular matrix (ECM), annular tears, development of disc herniation, and loss of disc height [4]. As a result of these anatomical changes, nerve root compression, spinal canal stenosis, and facet joint arthritis and hypertrophy can occur and can lead to chronic LBP and/or radiating leg pain with or without neurological deficits [5]. Moreover, the inflammatory environment of the degenerative discs and neurite sprouting have been suggested as the cause of discogenic LBP [6]. Thus, discogenic LBP is associated with complex interactions between the mechanical aspects of the IVD, inflammation, and the central or peripheral nervous system [7].

Chronic LBP due to IVDD may be considered for surgical treatments if there is no response to conservative treatments, such as medication and physical therapy. Surgical treatments include discectomy to remove a herniated disc, spinal fusion surgery used to connect two vertebrae to limit the movement of the spinal motion segment, and artificial disc replacements designed to restore and maintain range of motion [8,9]. However, spinal fusion surgery does not restore the previous range of motion and mechanical load-bearing properties of the IVD. Moreover, spinal fusion can lead to adjacent segment disease, which is a typical long-term complication after spinal fusion surgery and further indicates disc herniation, spinal canal stenosis, or spondylosis at levels above or below the index fusion level [10]. Therefore, alternate biological therapies are needed prior to surgery to slow or reverse the progression of IVDD, which usually leads to pain and disability.

The mechanical properties of the IVD are critical for proper functioning. In vivo, the IVD is the load-bearing structure of the spine and is subjected to spinal tension, torsion, compression, and bending [11]. Anatomically, the normal IVD consists of the following three parts: (1) the nucleus pulposus (NP), which contains a highly hydrated gel-like matrix comprising PGs and type II collagen; (2) the annulus fibrosus (AF), which is composed of lamellae in which parallel type I collagen fibers located within each lamella are aligned and help the IVD maintain its integrity from bending, stretching, and twisting; and (3) the end plates (EP), which consist of osseous and two hyaline cartilages [12] (Figure 1). In recent years, the understanding of IVD development, cell biology, and mechanisms of IVD degeneration has significantly advanced, enabling the development of biological approaches for IVD regeneration [4]. Biological approaches include growth factor injection and gene and cell-based therapies, whereas tissue engineering approaches involve the restoration of the mechanical and biological properties of the tissue via the addition of biomaterials to the degenerated disc [13].

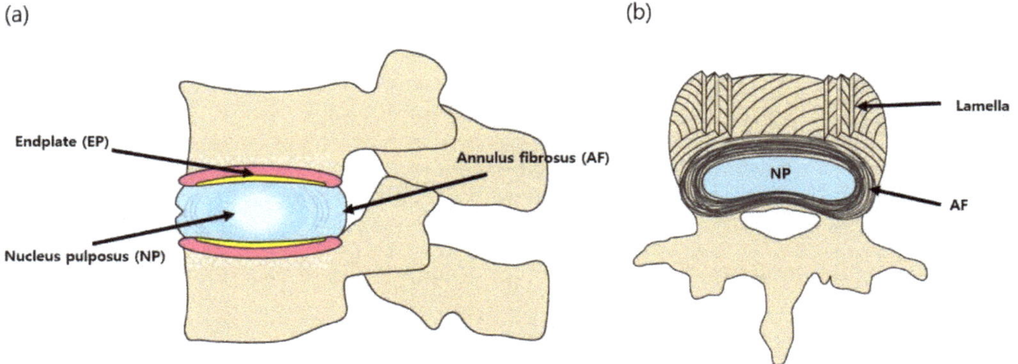

Figure 1. Schematic diagram of the intervertebral disc (IVD). (**a**) sagittal cross-section (**b**) axial cross-section.

Depending on the stage of IVDD, different treatment strategies for managing IVDD have been recommended [14]. Initial IVDD shows change in the NP and AF matrices, while IVDD induces progressive structural changes such as annular fissures, disc herniation, disc height reduction, and disc space collapse. The Pfirrmann Disc Grading is a useful scoring tool for evaluating IVDD on MRI T2-weighted images; Grade I: homogeneous disc with bright high intensity and normal disc height; Grade II: inhomogeneous disc but high intensity signal, clear distinction between NP and AF, and normal disc height; Grade III: inhomogeneous disc with an intermittent gray signal intensity, unclear distinction between NP and AF, and normal or slightly reduced disc height; Grade IV: inhomogeneous disc with low intensity dark gray signal intensity, no more distinction between the NP and AF, and slightly or moderately decreased disc height; Grade V: inhomogeneous disc with a hypointense black signal intensity, no more distinction between the nucleus and annulus, and collapsed disc space [15]. For severe nerve compression due to herniated disc and disc space collapse (Pfirrmann grades IV and V), surgical removal of herniated disc and/or spinal fusion surgery are required. However, patients with discogenic LBP due to Pfirrmann grade II and III IVDD may receive regenerative molecular therapies such as growth factors, genes, and cell therapy with or without biomaterials. In addition, tissue engineered NP and AF could be applied for patients with Pfirrmann grades IV and V.

The aim of this review was to provide an overview of the current status of tissue engineering applications for the treatment of IVD regeneration. To obtain an overview of the current tissue engineering strategies utilized for the repair of degenerated discs, we conducted a literature search using PubMed (https://pubmed.ncbi.nlm.nih.gov/, from 1 January 2000 and 1 December 2020) and the following key words: "intervertebral disc", "regeneration", "stem cells", "biomaterials", and "tissue engineering".

2. Biological Approaches

2.1. Molecular Therapies

2.1.1. Growth Factors

In the early stages of IVDD, growth factor injection may rebalance the anabolic and catabolic pathways in the degenerative cascade [16]. Degenerated discs have been reported to be repaired by intradiscal injection of growth factors, including insulin-like growth factor-1(IGF-1), epidermal growth factor (EGF), basic fibroblast growth factor (bFGF or FGF2), transforming growth factor-β (TGF-β), bone morphogenetic protein-2 (BMP-2), BMP-7 (osteogenic protein-1; OP-1), and growth and differentiation factor-5 (GDF-5) by promoting cell proliferation and matrix synthesis in experimental models [17–19]. The mitogenic potential of human NP and AF cells can be stimulated by platelet-derived growth factor (PDGF), bFGF, and IGF-1 [20,21]. PDGF, a known angiogenic growth factor, has been shown to inhibit IVD cell apoptosis and promote anabolic gene expression [22]. In several animal experiments, intradiscal injection of BMP-7 has shown improvements in the disc height and NP proteoglycan content [23,24]. GDF-5, also known as BMP-14, is another anabolic protein that promotes cell proliferation and proteoglycan synthesis in degenerated discs [18]. Similar to BMP-7, many animal studies of intradiscal injection of GDF-5 have shown improved disc height, cell proliferation, and matrix synthesis [25,26]. Despite their efficacy, there are many concerns surrounding the clinical use of recombinant growth factors due to their short half-life, limited stability, high cost, and problems associated with binding large molecules to polymers [27]. Thus, the right carrier is a matter to consider. In recent years, the use of biodegradable microspheres for controlled local drug delivery has become a valuable approach to overcome the drawbacks of growth factors. Yan et al. demonstrated that injection of GDF-5 loaded into poly(lactic-co-glycolic acid) (PLGA) microspheres could improve regenerative efficacy of GDF-5 in a rat model [28].

2.1.2. Gene Therapy

A strategy to overcome the short half-life limitation of growth factors is to provide a sustained supply of growth factors within the IVD [29]. The therapeutic effect of gene therapy is based on the induction of target gene upregulation or downregulation. These genes are transferred using viral or non-viral vectors, which are either directly injected into the degenerated discs or transduced into cells [16]. Another strategy used for intradiscal gene therapy is gene expression downregulation, which is detrimental to the physiological balance of the disc and may, thus, lead to IVDD [16]. Hence, if gene therapy is performed properly, it can provide many benefits, including a more sustained target gene expression and long-term biological effects. Promising targets for gene therapy include both anabolic regulators, such as TGF-β, latent membrane protein 1(LMP-1), and SOX-9, and anticatabolic regulators, such as anti-ADAMTS (a disintegrin and metalloproteinase with thrombospondin motifs)-5, and TIMP (tissue inhibitor of metalloproteinases)-1 [13]. In terms of non-virus vector-mediated gene transfer, ultrasound targeted microbubble transfection method has been reported to improve the transfection efficiency of plasmid DNA in NP cells, and polyplex micelles made from a vector carrying miRNA-25-3p were used in an IVDD rat model [30]. Clustered Regulatory Interspaced Short Palindromic

Repeats-Associated Cas9 (CRISPR/Cas9) is an innovative technology that can be used to target other genes for IVDD treatment. Using the CRISPR/Cas9 gene-editing system in AF cells from patients with chronic LBP, knock out of the transient receptor potential vanilloid type 4 (TRPV4) gene induced the reduction in inflammation [30]. However, the use of gene therapy for IVD regeneration is currently limited to in vitro and in vivo animal studies due to safety concerns. There are no ongoing human clinical trials of gene therapies for IVD regeneration.

2.1.3. Summary

In order to overcome the shortcomings of molecular therapy, future studies will focus on the delivery and controlled release to the degenerated discs. In addition, the combination of growth factors, stem cells, and biomaterials should be a focus going forward.

2.2. Cell-Based Therapies

In the intermediate stages of IVDD, cell transplantation can be used to repopulate the disc. Although the NP cell phenotype has not been well defined, the adult NP contains cells similar to chondrocyte [31]. Cell transplantation to a moderately degenerated disc is a possible treatment that promotes disc regeneration by reproliferating cells that can restore the structural and functional properties of the degenerated disc or delay degeneration [27].

An optimal source of cells suitable for cell transplantation in the degenerative disc remains elusive. Several studies have demonstrated that implantation of IVD-derived cells delays the process of progressive degeneration and, in some cases, promotes disc regeneration in an animal model of IVDD [32,33]. However, since these cells are derived from normal NP tissue, they cannot be obtained without damaging the IVD [25].

Stem cells are a promising candidate cell source for use in cell-based therapies for IVD repair. Most of the stem cells used in disc regeneration experiments are derived from the bone marrow, adipose tissue, umbilical cord blood, umbilical cord Wharton's jelly, and synovium, because these cells are relatively easy to obtain and can differentiate into chondrogenic and IVD-cell lineages [34,35]. Implantation of MSCs into degenerated discs can prevent cellular apoptosis and inflammation (paracrine effect) or differentiate MSCs into NP cells to restore normal homeostasis and prevent or reverse further degeneration [13]. Injecting stem cells into the degenerated disc has been reported to increase the proteoglycan and water contents of the disc ECM (Table 1) [29].

Notochordal cells have been suggested to cause disc degeneration because their loss is associated with the onset of IVDD [33]. Sheyn et al. demonstrated that notochordal-like cells from human induced pluripotent stem cells (iPSCs) reduce IVDD in an injury-induced porcine model [33]. Therefore, the application of iPSCs is a hot topic in the field of IVD regeneration and has several advantages over embryonic stem cells (ESCs), such as fewer ethics and immune rejection issues [33]. However, iPSCs also have drawbacks in clinical applications, such as tumor formation by genomic integration of reprogramming factors [36].

For the clinical application of stem cells for IVD regeneration, MSCs treatment strategies, cell doses, and efficacy are being investigated in various experimental settings of IVDD and clinical trials [37]. The use of MSCs is generally considered safe and effective in preventing IVDD, but the rate of osteophyte formation has been reported to be around 2.7% [37,38]. Osteophyte formation is believed to be the result of implanted MSC leakage. Therefore, the application of scaffolding materials, such as fibrin, hyaluronan, or atelocollagen, is strongly recommended to prevent cell leakage and reduce the risk of ectopic osteoblast differentiation of MSCs [39–50].

Table 1. Cell-based therapies for IVD regeneration.

Author	Cell Line	Effect
2019, Shi et al. [40] 2019, Sheyn et al. [33]	Neonatal human dermal fibroblasts Rabbit, in vivo Notochordal cells from human iPSCs	Increased regeneration markers Reduction of disc degeneration in a porcine model
2018, Teixeria et al. [41]	Human BM-MSC Bovine, ex-vivo	Promoted cell migration and increased inflammatory cytokine expression
2018, Wang et al. [42]	Rat BM-MSC Rat, in vivo	Hypoxic pre-treatment of BM-MSC with $CoCl_2$ enhanced migration, decreased apoptosis, increased disc height, MSC numbers in the NP and AF, and extracellular matrix production
2017, Maidhof et al. [43]	Allogeneic rat BM-MSC Rat, in vivo	Cell therapies administered at an early stage of injury or disease progression may have greater chances of mitigating matrix loss
2017, Hang et al. [44]	Autologous canine BM-MSC Canine, in vivo	PET was more reliable than MRI for quantifying implanted BM-MSC survival
2017, Steffen et al. [45]	Autologous canine BM-MSC Canine, in vivo	Successful injection of BM-MSC into lumbosacral discs of naturally IVD-degenerative canines
2017, Noriega et al. [46]	Allogeneic BM-MSC Clinical trial (N = 24, follow-up: 12 months)	Significant VAS and ODI reductions, improvement on MRI
2017, Centeno et al. [47]	Autologous BM-MSC Clinical trial (N = 33, follow-up: 72 months)	Disc bulging reduction on MRI, pain and function improvement
2017, Kumar et al. [48]	Autologous AD-MSC Clinical trial (N = 10, follow-up: 12 months)	Combined implantation of AD-MSC and hyaluronic acid in discogenic back pain is safe and tolerable
2017, Pettine et al. [49]	Autologous BM-MSC Clinical trial (N = 26, follow-up: 36 months)	Evidence for the safety and feasibility of intradiscal BM concentrate therapy
2016, Tschugg et al. [50]	Autologous disc chondrocyte Clinical trial (N = 120, follow-up: 48 months)	Ongoing study

BM-MSC, bone marrow-derived mesenchymal stem cell; AD-MSC, adipose tissue-derived mesenchymal stem cell; NP, nucleus pulposus; AF, annulus fibrosus; VAS, visual analogue scale; ODI, Oswestry disability index; iPSC, induced pluripotent stem cell; IVD, intervertebral disc.

To summarize, stem cell therapy can be used to induce IVD repair by preventing cellular apoptosis and inflammation, and by increasing the resident population and ECM production, and there is great interest in developing biomaterials for effective cell delivery, increasing cell viability, and inducing differentiation of stem cells into IVD-like cells.

3. Tissue Engineering for IVD Regeneration

Although many studies have reported alternative biological treatments for IVDD, these approaches have certain limitations. Direct administration of growth factors is associated with the short half-life of growth factors and potential lack of IVD cells as therapeutic targets in severe disc degeneration [29]. Gene therapy has several disadvantages, including inefficient gene delivery, unstable long-term expression, and lack of safety. While cell-based therapies have shown more promising therapeutic potential, the best strategies for effectiveness and safety have yet to be addressed [29]. In a clinical setting, stem cells are implanted in a harsh environment consisting of low cellularity, low glucose, low oxygen, low PH due to high lactic acid accumulation, low nutrients, and an inflammatory milieu [6,51–53]. Inflammatory mediators are a key component of progressive IVDD. All these factors can affect the differentiation potential, viability, and metabolism of the implanted stem cells [52]. MSCs can function optimally in inflammatory, hypoxic, acidic, and malnourished environments of the degenerated disc and have an immunomodulatory paracrine effect [52,54,55]. Hypoxia-exposed human MSCs (hMSCs) have been reported to improve tissue protection, but exposure of MSCs to inflammatory factors or hypoxic environment may adversely affect MSC differentiation. Therefore, it is important and necessary to design scaffolds for effective cell delivery and induction of stem cell differentiation for tissue engineering applications (Figure 2).

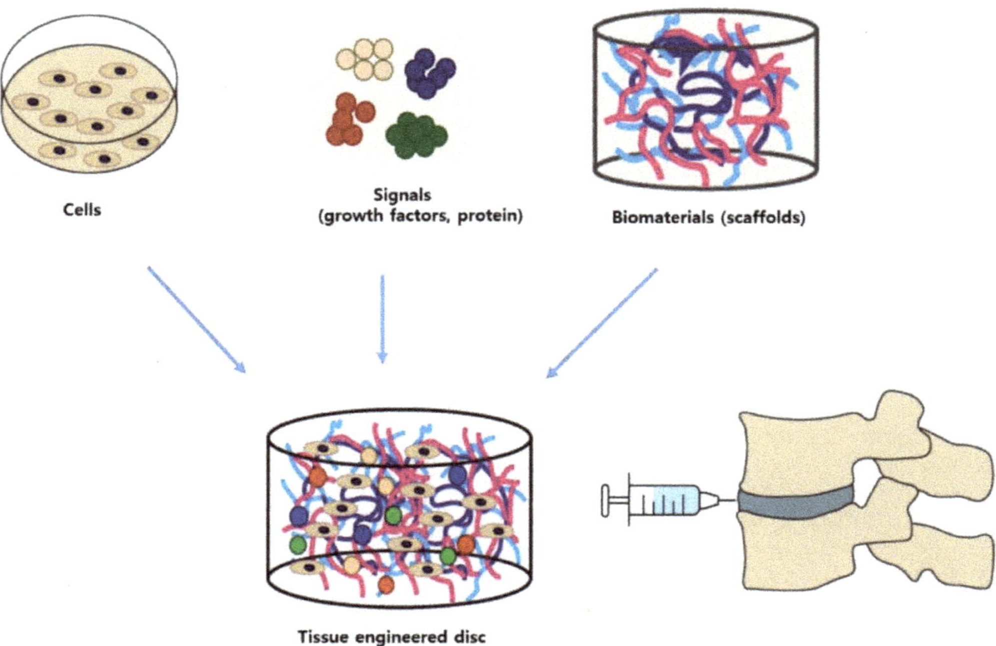

Figure 2. Combination of cells, signaling molecules (growth factors), and biomaterials for tissue engineering applications for intervertebral disc (IVD) degeneration.

3.1. Biomaterials

In the case of severe IVDD with loss of cell volume and of the physiological disc structure, the disc height must be restored to ensure IVD function [29]. This strategy involves a tissue engineering approach using biomaterials, which may serve as functional alternatives and scaffolds for the IVD tissue [13]. Researchers are using a composite approach that utilizes cell-loaded biomaterials to provide a structural environment for mechanical stability and potential cell regeneration [56]. The biomaterials used include injectable hydrogels and synthetic polymers designed from materials such as alginate, gelatin, polyglycolic acid, polylactic acid, hyaluronic acid (HA), and collagen [57].

Hydrogels can be used as an alternative to NP due to their biophysical properties and ability to absorb water, resist repeated loads, and act as a delivery vehicle [58]. An ideal injectable biomaterial will support cell retention and survival and make it possible to maintain or promote the NP phenotype in vivo. In the absence of biomaterials, cell injection leads to rapid cell death or migration from the injection site [59,60]. Important parameters to consider in the development of biomaterials are material viscosity, gelation rate, final gel stiffness, adhesivity, and degradation time controlled by polymer composition.

HA is a key component of the NP ECM that provides resistance to compression and allows for periodic loading [61,62]. Physiologically, HA has been demonstrated to have cartilage protective and anti-inflammatory properties, which have been shown to be associated with cell-based interventions. Therefore, some clinical trials have used HA as a cell carrier to increase the viscosity of the cell solution and enhance the retention of injected cells [63,64].

Collagen is one of the most widely used materials for tissue regeneration as it has numerous adhesion sites, limited immunogenicity, and is injectable. However, due to its poor degradation and mechanical properties, it has not been widely used for disc repair. Composite collagen hydrogels, on the other hand, have been found to improve the compressive mechanical properties of the scaffold and control the rate of scaffold degradation [27].

Fibrin is a naturally occurring biomaterial that provides intrinsic physical and soluble cues to initiate tissue repair. Biodegradable fibrin hydrogels can be produced as injectable cell carriers and can be mediated by adjusting coagulation protein levels or altering the ionic strength of the system [65,66]. Fibrin-only hydrogels remain vulnerable to cell-mediated remodeling, while fibrin-HA composite hydrogels improve stability by increasing glycosaminoglycan (GAG) synthesis. In addition, silk to fibrin-HA gels significantly improve the mechanical properties and promote chondrogenesis [67]. Silk offers high resistance to compression, and silk-fiber stability, due to its wide range of hydrogen bonds, protein hydrophobicity, and high crystallinity, provides an advantage as a scaffold for IVD bioengineering [68].

Other biomaterials that can be used as a matrix to support AF and NP engineering are chitosan and alginate, which are inexpensive and easily accessible [12]. In addition, these two polymers have a synergistic effect combined with hybrid scaffolding [69]. Chitosan is used as a biodegradable and biocompatible polymer with low toxicity and excellent antibacterial properties. The soft, spongy chitosan-based scaffold has high porosity and pore interconnectivity to support cell adhesion and growth [70]. Alginate, one of the most abundant natural materials, mainly derived from brown algae and some bacteria, is used in a variety of biomedical applications and drug delivery systems due to its excellent biocompatibility, biodegradability, non-antigenicity, and chelation ability [69].

The use of synthetic materials as injectable fillers or cell carriers is a promising strategy to prevent the biomechanical limitations of natural polymer-based hydrogels. Many synthetic biomaterials, such as polyethylene glycol (PEG), PLGA, polyvinyl alcohol (PVA), polyvinylpyrrolidone (PVP), and hydroxyethyl methacrylate (HEMA), have been used as drug delivery and cell carriers [71,72]. The first approach used to restore NP height, function, and motion focused on the use of in situ hydrated synthetic polymers to restore NP hydration, IVD pressure, and disc height. This tactic aimed at mimicking the hydration properties of the NP glycosaminoglycans, which are slowly degraded and modified as they age and degenerate [57]. A copolymeric hydrogel, with the longest history of clinical use, has served as an alternative to NP. However, similar to previously used in situ hydrated polymers, some complications have already been reported, including gel fragmentation during swelling [73]. Other biomaterials, such as NP implant devices, have been developed from an injectable polymer that physically transitions to a gel or solid form. This approach has the advantage of minimizing tissue damage to AF during transplantation. In addition, various strategies, such as chemical cross-linking agents and heat- or pH-induced transition, have been developed. The cross-linked material inhibits proteolysis and induces the stiffness for disc implant [57].

Many scientists have attempted to develop monophasic NP or AF scaffolds such as ECM-based scaffold [74], adipose-derived MSCs (AD-MSCs)-loaded NP tissue-engineered construct [75], ASCs-seeded type II collagen/chondroitin sulfate composite hydrogel [76], and decellularized NP-based scaffold [77] (Table 2). Extensive efforts have also been made to develop biphasic tissue-engineered AF-NP composite scaffolds such as a collagen-GAG co-precipitate (NP-like core)/photochemically crosslinked collagen membranes (AF-like lamellae) composite [78], multiple AF-like lamellae encasing an NP-like core (MSCs-seeded collagen-GAG) [79], engineered nanofibrous disk-like angle-ply structure [80], biomimetic glycosaminoglycan analogues based on sulphonate-containing polymers [81], and multiple HA-PEG composite hydrogel [82] (Table 2). Animal studies using these biphasic scaffolds have shown excellent mechanical and biochemical results, suggesting that mimicking the morphology of the IVD is important for long-term stability and function of the implanted scaffolds [78–82].

Table 2. Biomaterials for IVD regeneration.

Author	Materials	Effect
2020, Penolazzi et al. [74]	Decellularized Wharton's jelly matrix from human umbilical cord as ECM-based scaffold	Promoted cell differentiation toward a discogenic phenotype, positively affected the expression of regulators of IVD homeostasis
2019, Ishiguro et al. [75]	AD-MSC-Tissue engineered construct Rat	Regenerative efficacy was investigated structurally and biomechanically up to 6 months after implantation
2018, Zhou et al. [76]	Type II collagen/chondroitin sulfate (CS) composite hydrogel-like adipose-derived stem cell delivery system	Minimally invasive approach to promote the regeneration of degenerated NP
2018, Zhou et al. [77]	Injectable decellularized NP-based cell delivery system (NPCS)	The mechanical properties of the NPCS system were similar to those of fresh NP; Biocompatible; It induced NP-like differentiation and ECM synthesis
2015, Choy et al. [78]	Collagen-glycosaminoglycan (GAG) co-precipitate and multiple lamellae of a photo-crosslinked collagen membrane	A biphasic scaffold comprising 10 AF-like lamellae had the best mechanical performance and elastic compliance
2015, Chik et al. [79]	Collagen-GAG coprecipitate MSC and contracted collagen gel, MSC	Spinal motion segment tissue engineering. Provided a 3D model for studying tissue maturation and functional remodeling
2014, Martin et al. [80]	Electrospun poly scaffold with cell-seeded hydrogels and disc-like angle-ply structure	Optimized the design of functional disc replacement in vivo
2014, Sivan et al. [81]	Biomimetic GAG analogue based on sulphonate-containing polymer	Provided intrinsic swelling pressure which could maintain disc hydration and height
2014, Jeong et al. [82]	Hyaluronic acid-poly(ethylene glycol) composite hydrogel	Highest number of NP and AF cells on HA-PEG hydrogels from lower molecular weight HA

AD-MSC, adipose tissue-derived mesenchymal stem cell; GAG, glycosaminoglycan; HA, hyaluronic acid.

In addition, many studies have demonstrated that differentiation of stem cell into the IVD cell phenotype is promoted by various types of biomaterials; (1) natural biomaterials: collagen type II-chondroitin sulphate hydrogel, gelatin-HA methacrylate hydrogel, silk-protein-based multilayered angle-ply scaffold, chitosan-HA hydrogel, decellularized allogenic IVD, acellular porcine NP hydrogel, NP cell-derived acellular matrix, dextran chitosan and teleostean combined hydrogel, temperature sensitive hydrogel (chitosan-glycerophosphate), chitosan and alginate gel scaffold, alginate and chitosan hydrogels, and self-assembling peptides; (2) Synthetic biomaterials: poly(N-isopropylacrylamide (pNIPAM) hydrogel system, nanofibrous poly(l-lactide) (PLLA) scaffolds, and heparin-poly(ethylene argininylaspartate digylceride) (PEAD) conjugated vehicle; (3) Biosynthetic biomaterials: T1307-fibrinogen hydrogel, HA-pNIPAM hydrogel, and pentosan polysulfate (PEG-HA-PPS) hydrogel [83].

In summary, transplanted stem cells should survive, proliferate, and differentiate into NP-/AF-like cells. The combination of biomaterials and stem cells can provide an effective strategy to enhance effective cell delivery and stem cell differentiation capacity. Although a variety of biomaterials have been studied to investigate the effects of biomaterials on cell delivery and stem cell differentiation, few materials are currently available for clinical application due to the limitations of mechanical properties, immunogenicity, and uncontrollable deviations in inducing stem cells differentiation. In addition to the mechanical properties and biocompatibility of biomaterials, maintaining stem cells activity in a local niche and enhancing the ability of stem cells to differentiate into NP and AF cells facilitates the application of biomaterials in clinical practice [74–83].

3.2. Tissue Engineering for AF and NP Restoration and Maintenance

The aim of tissue engineering for IVD degeneration is the restoration and maintenance of both AF and NP anatomy and function. Tissue-engineered scaffolds must be able to withstand the physiological IVD loads and have excellent biocompatibility; proper porosity; and shapes, structures, and mechanical properties similar to those of IVD [79].

3.2.1. AF Regeneration and Tissue Engineering

AF is composed of type I collagen and stacked lamellae and is highly organized [84]. AF is needed to transfer stress from the NP, maintain IVD integrity, and protect against damage caused by bending, stretching, and twisting [85,86]. The homeobox protein Mohawk (Mkx) has been reported to be essential for AF development, maintenance, and regeneration. It has been found that Mkx is predominantly expressed in the outer AF, and that removal of Mkx in mice resulted in the loss of numerous tendon- and ligament-related genes in the outer AF. Transplantation of MSCs overexpressing Mkx revealed the AF phenotype and promoted functional AF regeneration [87].

In AF regeneration and tissue engineering, natural materials, such as collagen, HA, chitosan, alginate, silk fibroin, and chondroitin sulfate, as well as natural biologic materials, such as the decellularized matrix from AF, are used to promote tissue regeneration and repair [57,88]. Natural scaffolds have the advantages of having low toxicity and similar properties to those of native tissue, and they can be mass-produced. Synthetic polymer scaffolds can be manufactured and processed based on the desired structural (aligned, angle-ply, hierarchical, bilayer, or biphasic) and mechanical properties of the final engineered tissue [11]. A poly(trimethylene carbonate) (PTMC) scaffold covered with a poly(ester-urethane) (PU) membrane to address AF rupture repair of bovine IVD has been manufactured as a carrier for MSCs. A PTMC scaffold with MSCs and PU membrane has been found to restore the disc height and prevent IVD herniation [89]. Furthermore, biodegradable poly(ether carbonate urethane)urea (PECUU) materials have been produced in AF-derived stem cells (AFSCs) using an electrospinning technique. Moreover, it has been reported that the elasticity of PECUU fibrous scaffolds with AFSCs resembled that of natural AF tissue [90].

3.2.2. NP Regeneration and Tissue Engineering

NP is composed of type II collagen and PGs and contains 77% water. In recent years, bioengineered scaffolds that resemble the native NP structure and its mechanical properties have attracted attention [11]. Easy to inject high molecular weight hyaluronic acid-gelatin-adipic acid dehydrazide (oxi-HAG-ADH) hydrogels with anti-inflammatory and immunosuppressive activities, low viscosity, viscoelasticity similar to that of NP tissue, and expression of NP ECM genes have been fabricated [91]. Choi et al. generated hyaluronic-methylcellulose (HAMC) hydrogels loaded with Wharton's jelly-derived MSCs (WJ-MSCs), which significantly promoted degenerated disc repair by improving NP cell viability and decreasing ECM degradation [92]. Gan et al. generated a hydrogel with dextran and gelatin as the first network and PEG as the second network to produce hydrogels, forming the optimal 3D interpenetrating network hydrogel. This increased NP cell proliferation, long-term cell retention and survival, and promoted rehydration and regeneration of degenerative NP in animal models [93]. Laminin is the main component of the NP ECM and directly interacts with NP cells to regulate their function. Several laminin mimetic peptides bound to polyacrylamide gels have been reported to be able to support an immature and healthy NP phenotype. These hydrogel scaffolds provided a favorable environment for NP cell proliferation [94]. Wan et al. manufactured a biocompatible self-assembled peptide hydrogel (SAPH) with easily modifiable properties and nanofibrous architecture. They reported that the SAPH scaffold was as strong as native tissue, injectable, and that it restored the IVD cell phenotype and stimulated deposition of aggrecan and type II collagen, which are key NP ECM components [95].

3.2.3. NP-AF Regeneration and Tissue Engineering

NP-AF tissue engineering combines two approaches: NP replacement and AF repair. There are three ways to manufacture NP-AF, namely, using NP and AF cell-seeding scaffolds, integrated biphasic NP-AF scaffolds, and scaffolds made with decellularized natural IVD [11]. Scaffolds seeded with NP and AF cells were separately prepared and assembled into a composite construct. Nesti et al. manufactured a biphasic construct using electrospun MSCs seeded on a PLLA scaffold and HA hydrogel [96]. Choy et al. generated a biphasic NP-AF scaffold with integrated collagen and glycosaminoglycans. The biphasic scaffold was composed of collagen-glycosaminoglycan, which coprecipitates as an NP-like core, and encapsulated in multiple lamellae of photochemically cross-linked collagen membranes that made up AF-like lamellae [79]. This scaffold exhibited mechanical characteristics similar to those of native discs with a ring-independent height recovery of 82–89%. Park et al. generated a scaffold consisting of chondrocytes and AF cells, which were respectively seeded into a scaffold consisting of hydrogel in the center and silk protein in the periphery, respectively [97]. Yang et al. manufactured an IVD scaffold by inversely reconstructing the structure of native IVD and bioprinting bacterial cellulose nanofibers using a high-throughput-optimized micropattern screening microchip in rats [98]. Chan et al. made a 70% endogenous cell-removing scaffold that preserved the glycosaminoglycan content, collagen fibril structure, and mechanical properties of the IVD by altering chemical and physical decellularization [99]. Hensley et al. created a natural NP-AF composite scaffold using decellularized bovine tail IVD and confirmed the presence of type II collagen and glycosaminoglycan in the NP region and the native angle-ply collagen microarchitecture in the AF region [100].

3.2.4. Summary

Tissue engineering techniques have emerged as a possible approach to treat IVDD by replacing degenerated discs with appropriate stem cells and biomaterials. Tissue engineered AF and NP can restore their function by repairing or replacing degenerated discs. Therefore, considerable research is underway on the development of scaffolds suitable for AF and NP regeneration. Many natural and synthetic biomaterials can be used as supporting matrices in AF and NP scaffolds [12]. Advances in manufacturing technologies, material processing and development, surface functionalization, drug delivery systems, and cell integration have accelerated the development of tissue engineering therapies for IVDD.

4. Conclusions

Chronic LBP due to IVDD represents a significant health and social burden. Regenerative tactics are being investigated with significant advances in understanding the characteristics of IVDD (Table 3). Current promising strategies include growth factor injection, gene therapy, cell-based therapy, and tissue engineering using biomaterials. In this review, we investigated biological and tissue engineering approaches for the treatment of IVD degeneration and regeneration strategies. Limitations of biological approaches that remain to be overcome include the short half-life and possible lack of IVD endogenous cells associated with growth factor injection therapy, inefficient gene delivery, unstable long-term expression, and safety issues in gene therapy, and the inflammatory environment, low pH, low oxygen tension, and poor nutritional availability in cell-based therapies. Promising tissue engineering strategies using cells, growth factors, and biomaterials could be utilized to overcome these problems. With the development of tissue engineering, scaffolds are considered the 'holy grail' of IVD repair [11]. However, tissue engineering therapy remains challenging due to a lack of accurate understanding of the underlying molecular mechanisms and regulation of IVD physiology. To date, there are no FDA approved intradiscal therapies associated with tissue engineering therapy. Therefore, more sophisticated materials and strategies for clinical application need to be developed. In addition, accurate diagnosis of IVDD and evaluation of therapeutic effectiveness are critical to the development of successful biological therapies. Although T2 mapping and diffusion weighted images(DWI) are newly quantified methods for IVDD evaluation [101], the development of improved non-destructive imaging techniques is essential to evaluate IVDD.

Table 3. Type of regenerative therapies for intervertebral disc regeneration.

Type		Advantages	Disadvantages
Growth factor	GDF-5, IGF-1, TGF-β, bFGF, OP-1	Stimulation of ECM production	Short half-life Need repeated injection
Gene therapy	Virus mediated Non-virus mediated RNAi CRISPR/Cas9	Long-lasting and timeless effects	Safety concerns Ethical concerns Significant cost
Stem cell	ESCs	Differentiation into three germ layers Self-renewal and high replication	Immune rejection concern Ethical concerns Potential for tumor formation
	iPSCs	Less ethical concerns than ESCs Patient-specific Autologous	Need method standardization Potential for tumor formation Need validation for safety
	MSCs: Bone marrow, Adipose tissue, umbilical cord Wharton's jelly Synovial membrane	Less ethical concerns than ESCs and iPSCs	Less cell proliferation Limit differentiation potential
Tissue engineering	Combination: stem cells, biomaterials, and growth factors	Ideal constructs	Need validation for biodegradation, biocompatibility, and optimal

GDF-5, growth differentiation factor-5; IGF-1, insulin-like growth factor-1; TGF-β, transforming growth factor- β, bFGF, basic fibroblast growth factor, OP-1, osteogenic protein-1; ECM, extracellular matrix; RNAi, RNA interference; CRISPR/Cas9, Clustered Regulatory Interspaced Short Palindromic Repeats-Associated Cas9; ESCs, embryonic stem cells; iPSCs, induced pluripotent stem cells; MSCs, mesenchymal stem cells.

Author Contributions: Conceptualization and methodology: I.H.; writing: C.K.L., D.H.H. and I.H.; data acquisition: H.C. (Hungtae Chung), E.J.R., A.D., J.W.K., H.C. (Hyemin Choi), S.Y.K. and B.B. All authors have read and agreed to the published version of the manuscript.

Funding: This research was supported by the Korea Health Technology Research and Development Project, Ministry for Health and Welfare Affairs (HR16C0002, HI20C0579) and a grant of the National Research Foundation of Korea (NRF) (2020R1A2C4001870.

Institutional Review Board Statement: Not applicable.

Informed Consent Statement: Not applicable.

Conflicts of Interest: The authors declare no conflict of interest.

References

1. Coric, D.; Pettine, K.; Sumich, A.; Boltes, M.O. Prospective study of disc repair with allogeneic chondrocytes Presented at the 2012 Joint Spine Section Meeting. *J. Neurosurg. Spine* **2013**, *18*, 85–95. [CrossRef]
2. de Schepper, E.I.; Damen, J.; van Meurs, J.B.; Ginai, A.Z.; Popham, M.; Hofman, A.; Koes, B.W.; Bierma-Zeinstra, S.M. The association between lumbar disc degeneration and low back pain: The influence of age, gender, and individual radiographic features. *Spine* **2010**, *35*, 531–536. [CrossRef]
3. Murray, C.J.; Lopez, A.D. Measuring the global burden of disease. *N. Engl. J. Med.* **2013**, *369*, 448–457. [CrossRef]
4. Urban, J.P.; Roberts, S. Degeneration of the intervertebral disc. *Arthritis Res. Ther.* **2003**, *5*, 120–130. [CrossRef] [PubMed]
5. Hoy, D.; Bain, C.; Williams, G.; March, L.; Brooks, P.; Blyth, F.; Woolf, A.; Vos, T.; Buchbinder, R. A systematic review of the global prevalence of low back pain. *Arthritis Rheum.* **2012**, *64*, 2028–2037. [CrossRef] [PubMed]

6. Risbud, M.V.; Shapiro, I.M. Role of cytokines in intervertebral disc degeneration: Pain and disc content. *Nat. Rev. Rheumatol.* **2014**, *10*, 44–56. [CrossRef] [PubMed]
7. Lotz, J.C.; Ulrich, J.A. Innervation, inflammation, and hypermobility may characterize pathologic disc degeneration: Review of animal model data. *J. Bone Jt. Surg.* **2006**, *88* (Suppl. 2), 76–82. [CrossRef]
8. Phillips, F.M.; Slosar, P.J.; Youssef, J.A.; Andersson, G.; Papatheofanis, F. Lumbar spine fusion for chronic low back pain due to degenerative disc disease: A systematic review. *Spine* **2013**, *38*, E409–E422. [CrossRef]
9. Geisler, F.H.; McAfee, P.C.; Banco, R.J.; Blumenthal, S.L.; Guyer, R.D.; Holt, R.T.; Majd, M.E. Prospective, randomized, multicenter FDA IDE study of CHARITÉ artificial disc versus lumbar fusion: Effect at 5-year follow-up of prior surgery and prior discectomy on clinical outcomes following lumbar arthroplasty. *SAS J.* **2009**, *3*, 17–25. [CrossRef]
10. Ghiselli, G.; Wang, J.C.; Bhatia, N.N.; Hsu, W.K.; Dawson, E.G. Adjacent segment degeneration in the lumbar spine. *J. Bone Jt. Surg.* **2004**, *86*, 1497–1503. [CrossRef] [PubMed]
11. Zhao, R.; Liu, W.; Xia, T.; Yang, L. Disordered mechanical stress and tissue engineering therapies in intervertebral disc degeneration. *Polymers* **2019**, *11*, 1151. [CrossRef] [PubMed]
12. Stergar, J.; Gradisnik, L.; Velnar, T.; Maver, U. Intervertebral disc tissue engineering: A brief review. *Bosn. J. Basic Med. Sci.* **2019**, *19*, 130–137. [CrossRef]
13. Ju, D.G.; Kanim, L.E.; Bae, H.W. Intervertebral disc repair: Current concepts. *Glob. Spine J.* **2020**, *10* (Suppl. 2), 130S–136S. [CrossRef] [PubMed]
14. Wu, P.H.; Kim, H.S.; Jang, I.-T. Intervertebral disc diseases PART 2: A review of the current diagnostic and treatment strategies for intervertebral disc disease. *Int. J. Mol. Sci.* **2020**, *21*, 2135. [CrossRef]
15. Griffith, J.F.; Wang, Y.-X.J.; Antonio, G.E.; Choi, K.C.; Yu, A.; Ahuja, A.T.; Leung, P.C. Modified Pfirrmann Grading System for lumbar intervertebral disc degeneration. *Spine* **2007**, *32*, E708–E712. [CrossRef]
16. Dowdell, J.; Erwin, M.; Choma, T.; Vaccaro, A.; Iatridis, J.; Cho, S.K. intervertebral disk degeneration and repair. *Neurosurgery* **2017**, *80*, S46–S54. [CrossRef]
17. Travascio, F.; Elmasry, S.; Asfour, S. Modeling the role of IGF-1 on extracellular matrix biosynthesis and cellularity in intervertebral disc. *J. Biomech.* **2014**, *47*, 2269–2276. [CrossRef]
18. Feng, C.; Liu, H.; Yang, Y.; Huang, B.; Zhou, Y. Growth and differentiation factor-5 contributes to the structural and functional maintenance of the intervertebral disc. *Cell. Physiol. Biochem.* **2015**, *35*, 1–16. [CrossRef]
19. Cho, H.; Lee, S.; Park, S.H.; Huang, J.; Hasty, K.A.; Kim, S.J. Synergistic effect of combined growth factors in porcine intervertebral disc degeneration. *Connect. Tissue Res.* **2013**, *54*, 181–186. [CrossRef] [PubMed]
20. O'Connell, G.D.; Newman, I.B.; Carapezza, M.A. Effect of long-term osmotic loading culture on matrix synthesis from intervertebral disc cells. *BioRes. Open Access* **2014**, *3*, 242–249. [CrossRef]
21. Sampat, S.R.; O'Connell, G.D.; Fong, J.V.; Alegre-Aguarón, E.; Ateshian, G.A.; Hung, C.T. Growth factor priming of synovium-derived stem cells for cartilage tissue engineering. *Tissue Eng. Part A* **2011**, *17*, 2259–2265. [CrossRef]
22. Presciutti, S.M.; Paglia, D.N.; Karukonda, T.; Soung do, Y.; Guzzo, R.; Drissi, H.; Moss, I.L. PDGF-BB inhibits intervertebral disc cell apoptosis in vitro. *J. Orthop. Res.* **2014**, *32*, 1181–1188. [CrossRef] [PubMed]
23. Imai, Y.; Miyamoto, K.; An, H.S.; Thonar, E.J.-M.A.; Andersson, G.B.J.; Masuda, K. Recombinant human osteogenic protein-1 upregulates proteoglycan metabolism of human anulus fibrosus and nucleus pulposus cells. *Spine* **2007**, *32*, 1303–1309. [CrossRef] [PubMed]
24. Huang, K.Y.; Yan, J.J.; Hsieh, C.C.; Chang, M.S.; Lin, R.M. The in vivo biological effects of intradiscal recombinant human bone morphogenetic protein-2 on the injured intervertebral disc: An animal experiment. *Spine* **2007**, *32*, 1174–1180. [CrossRef]
25. Walsh, A.J.L.; Bradford, D.S.; Lotz, J.C. In vivo growth factor treatment of degenerated intervertebral discs. *Spine* **2004**, *29*, 156–163. [CrossRef] [PubMed]
26. Liang, H.; Ma, S.-Y.; Feng, G.; Shen, F.H.; Li, X.J. Therapeutic effects of adenovirus-mediated growth and differentiation factor-5 in a mice disc degeneration model induced by annulus needle puncture. *Spine J.* **2010**, *10*, 32–41. [CrossRef]
27. O'Connell, G.D.; Leach, J.K.; Klineberg, E.O. Tissue engineering a biological repair strategy for lumbar disc herniation. *BioRes. Open Access* **2015**, *4*, 431–445. [CrossRef]
28. Yan, J.; Yang, S.; Sun, H.; Guo, D.; Wu, B.; Ji, F.; Zhou, D. Effects of releasing recombinant human growth and differentiation factor-5 from poly(lactic-co-glycolic acid) microspheres for repair of the rat degenerated intervertebral disc. *J. Biomater. Appl.* **2014**, *29*, 72–80. [CrossRef] [PubMed]
29. Han, I.; Ropper, A.E.; Konya, D.; Kabatas, S.; Toktas, Z.; Aljuboori, Z.; Zeng, X.; Chi, J.H.; Zafonte, R.; Teng, Y.D. Biological approaches to treating intervertebral disk degeneration: Devising stem cell therapies. *Cell Transplant.* **2015**, *24*, 2197–2208. [CrossRef]
30. Roh, E.; Darai, A.; Kyung, J.; Choi, H.; Kwon, S.; Bhujel, B.; Kim, K.; Han, I. Genetic therapy for intervertebral disc degeneration. *Int. J. Mol. Sci.* **2021**, *22*, 1579. [CrossRef]
31. Risbud, M.V.; Schoepflin, Z.R.; Mwale, F.; Kandel, R.A.; Grad, S.; Iatridis, J.C.; Sakai, D.; Hoyland, J.A. Defining the phenotype of young healthy nucleus pulposus cells: Recommendations of the Spine Research Interest Group at the 2014 annual ORS meeting. *J. Orthop. Res.* **2015**, *33*, 283–293. [CrossRef]
32. Sakai, D.; Andersson, G.B.J. Stem cell therapy for intervertebral disc regeneration: Obstacles and solutions. *Nat. Rev. Rheumatol.* **2015**, *11*, 243–256. [CrossRef]

33. Sheyn, D.; Ben-David, S.; Tawackoli, W.; Zhou, Z.; Salehi, K.; Bez, M.; De Mel, S.; Chan, V.; Roth, J.; Avalos, P.; et al. Human iPSCs can be differentiated into notochordal cells that reduce intervertebral disc degeneration in a porcine model. *Theranostics* **2019**, *9*, 7506–7524. [CrossRef]
34. Chen, S.; Emery, S.E.; Pei, M. Coculture of synovium-derived stem cells and nucleus pulposus cells in serum-free defined medium with supplementation of transforming growth factor-beta1: A potential application of tissue-specific stem cells in disc regeneration. *Spine* **2009**, *34*, 1272–1280. [CrossRef] [PubMed]
35. Leckie, S.K.; Sowa, G.A.; Bechara, B.P.; Hartman, R.A.; Coelho, J.P.; Witt, W.T.; Dong, Q.D.; Bowman, B.W.; Bell, K.M.; Vo, N.V.; et al. Injection of human umbilical tissue–derived cells into the nucleus pulposus alters the course of intervertebral disc degeneration in vivo. *Spine J.* **2013**, *13*, 263–272. [CrossRef]
36. Chang, E.-A.; Jin, S.-W.; Nam, M.-H.; Kim, S.-D. human induced pluripotent stem cells: Clinical significance and applications in neurologic diseases. *J. Korean Neurosurg. Soc.* **2019**, *62*, 493–501. [CrossRef] [PubMed]
37. Yim, R.L.-H.; Lee, J.T.-Y.; Bow, C.H.; Meij, B.; Leung, V.; Cheung, K.M.; Vavken, P.; Samartzis, D. A Systematic review of the safety and efficacy of mesenchymal stem cells for disc degeneration: Insights and future directions for regenerative therapeutics. *Stem Cells Dev.* **2014**, *23*, 2553–2567. [CrossRef]
38. Salzig, D.; Schmiermund, A.; Gebauer, E.; Fuchsbauer, H.-L.; Czermak, P. Influence of porcine intervertebral disc matrix on stem cell differentiation. *J. Funct. Biomater.* **2011**, *2*, 155–172. [CrossRef] [PubMed]
39. Vadalà, G.; Sowa, G.; Hubert, M.; Gilbertson, L.G.; Denaro, V.; Kang, J.D. Mesenchymal stem cells injection in degenerated intervertebral disc: Cell leakage may induce osteophyte formation. *J. Tissue Eng. Regen. Med.* **2011**, *6*, 348–355. [CrossRef]
40. Shi, P.; Chee, A.; Liu, W.; Chou, P.-H.; Zhu, J.; An, H.S. Therapeutic effects of cell therapy with neonatal human dermal fibroblasts and rabbit dermal fibroblasts on disc degeneration and inflammation. *Spine J.* **2019**, *19*, 171–181. [CrossRef] [PubMed]
41. Teixeira, G.Q.; Pereira, C.L.; Ferreira, J.R.; Maia, A.F.; Gomez-Lazaro, M.; Barbosa, M.A.; Neidlinger-Wilke, C.; Goncalves, R.M. Immunomodulation of human mesenchymal stem/stromal cells in intervertebral disc degeneration: Insights from a proinflammatory/degenerative ex vivo model. *Spine* **2018**, *43*, e673–e682. [CrossRef] [PubMed]
42. Wang, W.; Wang, Y.; Deng, G.; Ma, J.; Huang, X.; Yu, J.; Xi, Y.; Ye, X. Transplantation of hypoxic-preconditioned bone mesenchymal stem cells retards intervertebral disc degeneration via enhancing implanted cell survival and migration in rats. *Stem Cells Int.* **2018**, *2018*, 1–13. [CrossRef] [PubMed]
43. Maidhof, R.; Rafiuddin, A.; Chowdhury, F.; Jacobsen, T.; Chahine, N.O. Timing of mesenchymal stem cell delivery impacts the fate and therapeutic potential in intervertebral disc repair. *J. Orthop. Res.* **2017**, *35*, 32–40. [CrossRef]
44. Hang, D.; Li, F.; Che, W.; Wu, X.; Wan, Y.; Wang, J.; Zheng, Y. One-stage positron emission tomography and magnetic resonance imaging to assess mesenchymal stem cell survival in a canine model of intervertebral disc degeneration. *Stem Cells Dev.* **2017**, *26*, 1334–1343. [CrossRef] [PubMed]
45. Steffen, F.; Smolders, L.A.; Roentgen, A.M.; Bertolo, A.; Stoyanov, J. Bone marrow-derived mesenchymal stem cells as autologous therapy in dogs with naturally occurring intervertebral disc disease: Feasibility, safety, and preliminary results. *Tissue Eng. Part C Methods* **2017**, *23*, 643–651. [CrossRef] [PubMed]
46. Noriega, D.C.; Ardura, F.; Hernández-Ramajo, R.; Martín-Ferrero, M.; Sánchez-Lite, I.; Toribio, B.; Alberca, M.; García, V.; Moraleda, J.M.; Sánchez, A.; et al. Intervertebral disc repair by allogeneic mesenchymal bone marrow cells: A randomized controlled trial. *Transplantation* **2017**, *101*, 1945–1951. [CrossRef]
47. Centeno, C.; Markle, J.; Dodson, E.; Stemper, I.; Williams, C.J.; Hyzy, M.; Ichim, T.; Freeman, M. Treatment of lumbar degenerative disc disease-associated radicular pain with culture-expanded autologous mesenchymal stem cells: A pilot study on safety and efficacy. *J. Transl. Med.* **2017**, *15*, 1–12. [CrossRef]
48. Kumar, H.; Ha, D.-H.; Lee, E.-J.; Park, J.H.; Shim, J.H.; Ahn, T.-K.; Kim, K.-T.; Ropper, A.E.; Sohn, S.; Kim, C.-H.; et al. Safety and tolerability of intradiscal implantation of combined autologous adipose-derived mesenchymal stem cells and hyaluronic acid in patients with chronic discogenic low back pain: 1-year follow-up of a phase I study. *Stem Cell Res. Ther.* **2017**, *8*, 1–14. [CrossRef]
49. Pettine, K.A.; Suzuki, R.K.; Sand, T.T.; Murphy, M.B. Autologous bone marrow concentrate intradiscal injection for the treatment of degenerative disc disease with three-year follow-up. *Int. Orthop.* **2017**, *41*, 2097–2103. [CrossRef]
50. Tschugg, A.; Michnacs, F.; Strowitzki, M.; Meisel, H.J.; Thomé, C. A prospective multicenter phase I/II clinical trial to evaluate safety and efficacy of NOVOCART Disc plus autologous disc chondrocyte transplantation in the treatment of nucleotomized and degenerative lumbar disc to avoid secondary disease: Study protocol for a randomized controlled trial. *Trials* **2016**, *17*, 1–10. [CrossRef]
51. Grunhagen, T.; Shirazi-Adl, A.; Fairbank, J.C.; Urban, J.P. Intervertebral disk nutrition: A review of factors influencing concentrations of nutrients and metabolites. *Orthop. Clin. N. Am.* **2011**, *42*, 465–477. [CrossRef]
52. Krock, E.; Rosenzweig, D.H.; Haglund, L. The inflammatory milieu of the degenerate disc: Is mesenchymal stem cell-based therapy for intervertebral disc repair a feasible approach? *Curr. Stem Cell Res. Ther.* **2015**, *10*, 317–328. [CrossRef]
53. Wuertz, K.; Haglund, L. Inflammatory mediators in intervertebral disk degeneration and discogenic pain. *Glob. Spine J.* **2013**, *3*, 175–184. [CrossRef] [PubMed]
54. Alkhatib, B.; Rosenzweig, D.H.; Krock, E.; Roughley, P.J.; Beckman, L.; Steffen, T.; Weber, M.H.; Ouellet, J.A.; Haglund, L. Acute mechanical injury of the human intervertebral disc: Link to degeneration and pain. *Eur. Cells Mater.* **2014**, *28*, 98–111. [CrossRef]

55. Binch, A.L.A.; Cole, A.A.; Breakwell, L.M.; Michael, A.L.R.; Chiverton, N.; Cross, A.K.; Le Maitre, C.L. Expression and regulation of neurotrophic and angiogenic factors during human intervertebral disc degeneration. *Arthritis Res. Ther.* **2014**, *16*, 1–15. [CrossRef] [PubMed]
56. Fernandez-Moure, J.; Moore, C.A.; Kim, K.; Karim, A.; Smith, K.; Barbosa, Z.; Van Eps, J.; Rameshwar, P.; Weiner, B. Novel therapeutic strategies for degenerative disc disease: Review of cell biology and intervertebral disc cell therapy. *SAGE Open Med.* **2018**, *6*, 2050312118761674. [CrossRef]
57. Bowles, R.D.; Setton, L.A. Biomaterials for intervertebral disc regeneration and repair. *Biomaterials* **2017**, *129*, 54–67. [CrossRef]
58. Liang, C.-Z.; Li, H.; Tao, Y.-Q.; Peng, L.-H.; Gao, J.-Q.; Wu, J.-J.; Li, F.-C.; Hua, J.-M.; Chen, Q.-X. Dual release of dexamethasone and TGF-β3 from polymeric microspheres for stem cell matrix accumulation in a rat disc degeneration model. *Acta Biomater.* **2013**, *9*, 9423–9433. [CrossRef]
59. Francisco, A.T.; Mancino, R.J.; Bowles, R.D.; Brunger, J.M.; Tainter, D.M.; Chen, Y.-T.; Richardson, W.J.; Guilak, F.; Setton, L.A. Injectable laminin-functionalized hydrogel for nucleus pulposus regeneration. *Biomaterials* **2013**, *34*, 7381–7388. [CrossRef]
60. Henriksson, H.B.; Svanvik, T.; Jonsson, M.; Hagman, M.; Horn, M.; Lindahl, A.; Brisby, H. Transplantation of human mesenchymal stems cells into intervertebral discs in a xenogeneic porcine model. *Spine* **2009**, *34*, 141–148. [CrossRef] [PubMed]
61. Leckie, A.E.; Akens, M.K.; Woodhouse, K.A.; Yee, A.J.; Whyne, C.M. Evaluation of thiol-modified hyaluronan and elastin-like polypeptide composite augmentation in early-stage disc degeneration: Comparing 2 minimally invasive techniques. *Spine* **2012**, *37*, E1296–E1303. [CrossRef]
62. Malhotra, N.R.; Han, W.M.; Beckstein, J.; Cloyd, J.; Chen, W.; Elliott, D.M. An injectable nucleus pulposus implant restores compressive range of motion in the ovine disc. *Spine* **2012**, *37*, E1099–E1105. [CrossRef]
63. Gupta, P.K.; Chullikana, A.; Rengasamy, M.; Shetty, N.; Pandey, V.; Agarwal, V.; Wagh, S.Y.; Vellotare, P.K.; Damodaran, D.; Viswanathan, P.; et al. Efficacy and safety of adult human bone marrow-derived, cultured, pooled, allogeneic mesenchymal stromal cells (Stempeucel®): Preclinical and clinical trial in osteoarthritis of the knee joint. *Arthritis Res. Ther.* **2016**, *18*, 1–18. [CrossRef]
64. Park, Y.-B.; Ha, C.-W.; Lee, C.-H.; Yoon, Y.C. Cartilage regeneration in osteoarthritic patients by a composite of allogeneic umbilical cord blood-derived mesenchymal stem cells and hyaluronate hydrogel: Results from a clinical trial for safety and proof-of-concept with 7 years of extended follow-up. *Stem Cells Transl. Med.* **2017**, *6*, 613–621. [CrossRef] [PubMed]
65. Ahmed, T.A.; Dare, E.V.; Hincke, M. Fibrin: A versatile scaffold for tissue engineering applications. *Tissue Eng. Part B Rev.* **2008**, *14*, 199–215. [CrossRef] [PubMed]
66. Davis, H.; Miller, S.; Case, E.; Leach, J. Supplementation of fibrin gels with sodium chloride enhances physical properties and ensuing osteogenic response. *Acta Biomater.* **2011**, *7*, 691–699. [CrossRef]
67. Park, S.-H.; Cho, H.; Gil, E.S.; Mandal, B.B.; Min, B.-H.; Kaplan, D.L. Silk-fibrin/hyaluronic acid composite gels for nucleus pulposus tissue regeneration. *Tissue Eng. Part A* **2011**, *17*, 2999–3009. [CrossRef] [PubMed]
68. Park, S.-H.; Gil, E.S.; Mandal, B.B.; Cho, H.S.; Kluge, J.A.; Min, B.-H.; Kaplan, D.L. Annulus fibrosus tissue engineering using lamellar silk scaffolds. *J. Tissue Eng. Regen. Med.* **2012**, *6* (Suppl. 3), s24–s33. [CrossRef] [PubMed]
69. Li, Z.; Ramay, H.R.; Hauch, K.D.; Xiao, D.; Zhang, M. Chitosan–alginate hybrid scaffolds for bone tissue engineering. *Biomaterials* **2005**, *26*, 3919–3928. [CrossRef]
70. Kim, I.-Y.; Seo, S.-J.; Moon, H.-S.; Yoo, M.-K.; Park, I.-Y.; Kim, B.-C.; Cho, C.-S. Chitosan and its derivatives for tissue engineering applications. *Biotechnol. Adv.* **2008**, *26*, 1–21. [CrossRef] [PubMed]
71. Kranenburg, H.-J.C.; Meij, B.P.; Onis, D.; Van Der Veen, A.J.; Saralidze, K.; Smolders, L.A.; Huizinga, J.G.; Knetsch, M.L.W.; Luijten, P.R.; Visser, F.; et al. Design, synthesis, imaging, and biomechanics of a softness-gradient hydrogel nucleus pulposus prosthesis in a canine lumbar spine model. *J. Biomed. Mater. Res. Part B Appl. Biomater.* **2012**, *100*, 2148–2155. [CrossRef] [PubMed]
72. Kumar, D.; Gerges, I.; Tamplenizza, M.; Lenardi, C.; Forsyth, N.R.; Liu, Y. Three-dimensional hypoxic culture of human mesenchymal stem cells encapsulated in a photocurable, biodegradable polymer hydrogel: A potential injectable cellular product for nucleus pulposus regeneration. *Acta Biomater.* **2014**, *10*, 3463–3474. [CrossRef]
73. Durdag, E.; Ayden, O.; Albayrak, S.; Atci, I.B.; Armagan, E. Fragmentation to epidural space: First documented complication of Gelstix(TM.). *Turk. Neurosurg.* **2014**, *24*, 602–605.
74. Penolazzi, L.; Pozzobon, M.; Bergamin, L.S.; D'Agostino, S.; Francescato, R.; Bonaccorsi, G.; De Bonis, P.; Cavallo, M.; Lambertini, E.; Piva, R. Extracellular matrix from decellularized Wharton's jelly improves the behavior of cells from degenerated intervertebral disc. *Front. Bioeng. Biotechnol.* **2020**, *8*, 262. [CrossRef] [PubMed]
75. Ishiguro, H.; Kaito, T.; Yarimitsu, S.; Hashimoto, K.; Okada, R.; Kushioka, J.; Chijimatsu, R.; Takenaka, S.; Makino, T.; Sakai, Y.; et al. Intervertebral disc regeneration with an adipose mesenchymal stem cell-derived tissue-engineered construct in a rat nucleotomy model. *Acta Biomater.* **2019**, *87*, 118–129. [CrossRef] [PubMed]
76. Zhou, X.; Wang, J.; Fang, W.; Tao, Y.; Zhao, T.; Xia, K.; Liang, C.; Hua, J.; Li, F.; Chen, Q. Genipin cross-linked type II collagen/chondroitin sulfate composite hydrogel-like cell delivery system induces differentiation of adipose-derived stem cells and regenerates degenerated nucleus pulposus. *Acta Biomater.* **2018**, *71*, 496–509. [CrossRef] [PubMed]
77. Zhou, X.; Wang, J.; Huang, X.; Fang, W.; Tao, Y.; Zhao, T.; Liang, C.; Hua, J.; Chen, Q.; Li, F. Injectable decellularized nucleus pulposus-based cell delivery system for differentiation of adipose-derived stem cells and nucleus pulposus regeneration. *Acta Biomater.* **2018**, *81*, 115–128. [CrossRef]

78. Choy, A.T.H.; Chan, B.P. A Structurally and Functionally Biomimetic Biphasic Scaffold for Intervertebral Disc Tissue Engineering. *PLoS ONE* **2015**, *10*, e0131827. [CrossRef] [PubMed]
79. Chik, T.K.; Chooi, W.H.; Li, Y.Y.; Ho, F.C.; Cheng, H.W.; Choy, T.H.; Sze, K.Y.; Luk, K.K.D.; Cheung, K.M.C.; Chan, B.P. Bioengineering a Multicomponent Spinal Motion Segment Construct-A 3D Model for complex tissue engineering. *Adv. Healthc. Mater.* **2014**, *4*, 99–112. [CrossRef]
80. Martin, J.T.; Milby, A.H.; Chiaro, J.A.; Kim, D.H.; Hebela, N.M.; Smith, L.J.; Elliott, D.M.; Mauck, R.L. Translation of an engineered nanofibrous disc-like angle-ply structure for intervertebral disc replacement in a small animal model. *Acta Biomater.* **2014**, *10*, 2473–2481. [CrossRef] [PubMed]
81. Sivan, S.; Roberts, S.; Urban, J.; Menage, J.; Bramhill, J.; Campbell, D.; Franklin, V.; Lydon, F.; Merkher, Y.; Maroudas, A.; et al. Injectable hydrogels with high fixed charge density and swelling pressure for nucleus pulposus repair: Biomimetic glycosaminoglycan analogues. *Acta Biomater.* **2014**, *10*, 1124–1133. [CrossRef] [PubMed]
82. Jeong, C.G.; Francisco, A.T.; Niu, Z.; Mancino, R.L.; Craig, S.L.; Setton, L.A. Screening of hyaluronic acid–poly(ethylene glycol) composite hydrogels to support intervertebral disc cell biosynthesis using artificial neural network analysis. *Acta Biomater.* **2014**, *10*, 3421–3430. [CrossRef] [PubMed]
83. Peng, Y.; Huang, D.; Liu, S.; Li, J.; Qing, X.; Shao, Z. Biomaterials-induced stem cells specific differentiation into intervertebral disc lineage cells. *Front. Bioeng. Biotechnol.* **2020**, *8*, 56. [CrossRef]
84. Ishihara, H.; Warensjo, K.; Roberts, S.; Urban, J.P. Proteoglycan synthesis in the intervertebral disk nucleus: The role of extracellular osmolality. *Am. J. Physiol.* **1997**, *272 Pt 1*, C1499–C1506. [CrossRef]
85. Schollum, M.L.; Robertson, P.A.; Broom, N.D. ISSLS prize winner: Microstructure and mechanical disruption of the lumbar disc annulus: Part I: A microscopic investigation of the translamellar bridging network. *Spine* **2008**, *33*, 2702–2710. [CrossRef] [PubMed]
86. Yu, J.; Fairbank, J.C.T.; Roberts, S.; Urban, J.P.G. The elastic fiber network of the anulus fibrosus of the normal and scoliotic human intervertebral disc. *Spine* **2005**, *30*, 1815–1820. [CrossRef]
87. Nakamichi, R.; Ito, Y.; Inui, M.; Onizuka, N.; Kayama, T.; Kataoka, K.; Suzuki, H.; Mori, M.; Inagawa, M.; Ichinose, S.; et al. Mohawk promotes the maintenance and regeneration of the outer annulus fibrosus of intervertebral discs. *Nat. Commun.* **2016**, *7*, 12503. [CrossRef] [PubMed]
88. Xu, H.; Xu, B.; Yang, Q.; Li, X.; Ma, X.; Xia, Q.; Zhang, Y.; Zhang, C.; Wu, Y.; Zhang, Y. Comparison of decellularization protocols for preparing a decellularized porcine annulus fibrosus scaffold. *PLoS ONE* **2014**, *9*, e86723. [CrossRef] [PubMed]
89. Pirvu, T.; Blanquer, S.B.; Benneker, L.M.; Grijpma, D.W.; Richards, R.G.; Alini, M.; Eglin, D.; Grad, S.; Li, Z. A combined biomaterial and cellular approach for annulus fibrosus rupture repair. *Biomaterials* **2015**, *42*, 11–19. [CrossRef]
90. Zhu, C.; Li, J.; Liu, C.; Zhou, P.; Yang, H.; Li, B. Modulation of the gene expression of annulus fibrosus-derived stem cells using poly(ether carbonate urethane)urea scaffolds of tunable elasticity. *Acta Biomater.* **2016**, *29*, 228–238. [CrossRef] [PubMed]
91. Chen, Y.-C.; Su, W.-Y.; Yang, S.-H.; Gefen, A.; Lin, F.-H. In situ forming hydrogels composed of oxidized high molecular weight hyaluronic acid and gelatin for nucleus pulposus regeneration. *Acta Biomater.* **2013**, *9*, 5181–5193. [CrossRef]
92. Choi, U.Y.; Joshi, H.P.; Payne, S.; Kim, K.T.; Kyung, J.W.; Choi, H.; Cooke, M.J.; Kwon, S.Y.; Roh, E.J.; Sohn, S.; et al. An Injectable Hyaluronan–Methylcellulose (HAMC) Hydrogel combined with Wharton's jelly-derived mesenchymal Stromal cells (WJ-MSCs) promotes degenerative disc repair. *Int. J. Mol. Sci.* **2020**, *21*, 7391. [CrossRef] [PubMed]
93. Gan, Y.; Li, P.; Wang, L.; Mo, X.; Song, L.; Xu, Y.; Zhao, C.; Ouyang, B.; Tu, B.; Luo, L.; et al. An interpenetrating network-strengthened and toughened hydrogel that supports cell-based nucleus pulposus regeneration. *Biomaterials* **2017**, *136*, 12–28. [CrossRef]
94. Bridgen, D.T.; Fearing, B.V.; Jing, L.; Sanchez-Adams, J.; Cohan, M.C.; Guilak, F.; Chen, J.; Setton, L.A. Regulation of human nucleus pulposus cells by peptide-coupled substrates. *Acta Biomater.* **2017**, *55*, 100–108. [CrossRef]
95. Wan, S.; Borland, S.; Richardson, S.M.; Merry, C.L.; Saiani, A.; Gough, J.E. Self-assembling peptide hydrogel for intervertebral disc tissue engineering. *Acta Biomater.* **2016**, *46*, 29–40. [CrossRef] [PubMed]
96. Nesti, L.J.; Li, W.-J.; Shanti, R.M.; Jiang, Y.J.; Jackson, W.; Freedman, B.A.; Kuklo, T.R.; Giuliani, J.R.; Tuan, R.S. Intervertebral disc tissue engineering using a Novel Hyaluronic Acid–Nanofibrous Scaffold (HANFS) amalgam. *Tissue Eng. Part A* **2008**, *14*, 1527–1537. [CrossRef]
97. Park, S.-H.; Gil, E.S.; Cho, H.; Mandal, B.B.; Tien, L.W.; Min, B.-H.; Kaplan, D.L. Intervertebral disk tissue engineering using biphasic silk composite scaffolds. *Tissue Eng. Part A* **2012**, *18*, 447–458. [CrossRef]
98. Yang, J.; Wang, L.; Zhang, W.; Sun, Z.; Li, Y.; Yang, M.; Zeng, D.; Peng, B.; Zheng, W.; Jiang, X.; et al. Reverse reconstruction and bioprinting of bacterial cellulose-based functional total intervertebral disc for therapeutic implantation. *Small* **2018**, *14*, 1702582. [CrossRef] [PubMed]
99. Chan, L.K.; Leung, V.Y.; Tam, V.; Lu, W.W.; Sze, K.; Cheung, K.M. Decellularized bovine intervertebral disc as a natural scaffold for xenogenic cell studies. *Acta Biomater.* **2013**, *9*, 5262–5272. [CrossRef] [PubMed]
100. Hensley, A.; Rames, J.; Casler, V.; Rood, C.; Walters, J.; Fernandez, C.; Gill, S.; Mercuri, J.J. Decellularization and characterization of a whole intervertebral disk xenograft scaffold. *J. Biomed. Mater. Res. Part A* **2018**, *106*, 2412–2423. [CrossRef]
101. Da Costa, R.C.; De Decker, S.; Lewis, M.J.; Volk, H.; Canine Spinal Cord Injury Consortium (CANSORT-SCI). Diagnostic imaging in intervertebral disc disease. *Front. Vet. Sci.* **2020**, *7*, 588338. [PubMed]

Review

The Role of Hyaluronic Acid in Intervertebral Disc Regeneration

Zepur Kazezian [1,†], Kieran Joyce [1,2] and Abhay Pandit [1,*]

[1] CÚRAM, SFI Research Centre for Medical Devices, National University of Ireland Galway, H91 W2TY Galway, Ireland; z.kazezian@imperial.ac.uk (Z.K.); k.joyce10@nuigalway.ie (K.J.)
[2] School of Medicine, National University of Ireland Galway, H91 TK33 Galway, Ireland
* Correspondence: abhay.pandit@nuigalway.ie
† Zepur Kazezian is currently at Imperial College London, London SW7 2AZ, UK.

Received: 17 August 2020; Accepted: 7 September 2020; Published: 9 September 2020

Abstract: Intervertebral disc (IVD) degeneration is a leading cause of low back pain worldwide, incurring a significant burden on the healthcare system and society. IVD degeneration is characterized by an abnormal cell-mediated response leading to the stimulation of different catabolic biomarkers and activation of signalling pathways. In the last few decades, hyaluronic acid (HA), which has been broadly used in tissue-engineering, has popularised due to its anti-inflammatory, analgesic and extracellular matrix enhancing properties. Hence, there is expressed interest in treating the IVD using different HA compositions. An ideal HA-based biomaterial needs to be compatible and supportive of the disc microenvironment in general and inhibit inflammation and downstream cascades leading to the innervation, vascularisation and pain sensation in particular. High molecular weight hyaluronic acid (HMW HA) and HA-based biomaterials used as therapeutic delivery platforms have been trialled in preclinical models and clinical trials. In this paper, we reviewed a series of studies focused on assessing the effect of different compositions of HA as a therapeutic, targeting IVD degeneration. Overall, tremendous advances have been made towards an optimal form of a HA biomaterial to target specific biomarkers associated with IVD degeneration, but further optimization is necessary to address regeneration.

Keywords: hyaluronic acid; disc repair; annulus fibrosus repair; biomaterials; inflammatory biomarkers and signalling pathways

---

1. Introduction

A total of 80% of the world population suffer from low back pain (LBP) which is considered the most significant cause of disability, resulting in a negative socioeconomic impact [1,2]. In terms of disability-adjusted life years, LBP incurs a substantial burden over other health-related conditions [3,4]. Although LBP is more common in patients above 65 years old [5,6], it can start as early as in the late teenage years [7]. It is classified as the most expensive healthcare treatment, ranging between 12 to 90 billion dollars in the United States alone [8] and above 500 million pounds in the United Kingdom according to the nice guideline [9].

LBP is frequently associated with the deterioration of the intervertebral disc (IVD) due to abnormal cell-mediated response leading to the stimulation of different catabolic enzymes [10,11] and signalling pathways [12], imbalance in extracellular matrix composition and overexpression of the extracellular matrix-degrading enzymes [13]. The changes in the biomechanical elements, which are represented by unbalanced mechanical loading [14], genetic background [15] and reduced cellular activity, lead to continuous structural failure (Figure 1) [16]. In terms of biochemical changes, the deficiency in nutrient diffusion into the disc due to the calcified cartilaginous endplate (CEP) is a fundamental cause of IVD degeneration [17]. Additionally, increased elastin concentrations in the annulus fibrosus (AF) [18], and augmented proteoglycan breakdown in the nucleus pulposus (NP) cause a dramatic decrease in disc height and ultimately lead to biomechanical instability [19].

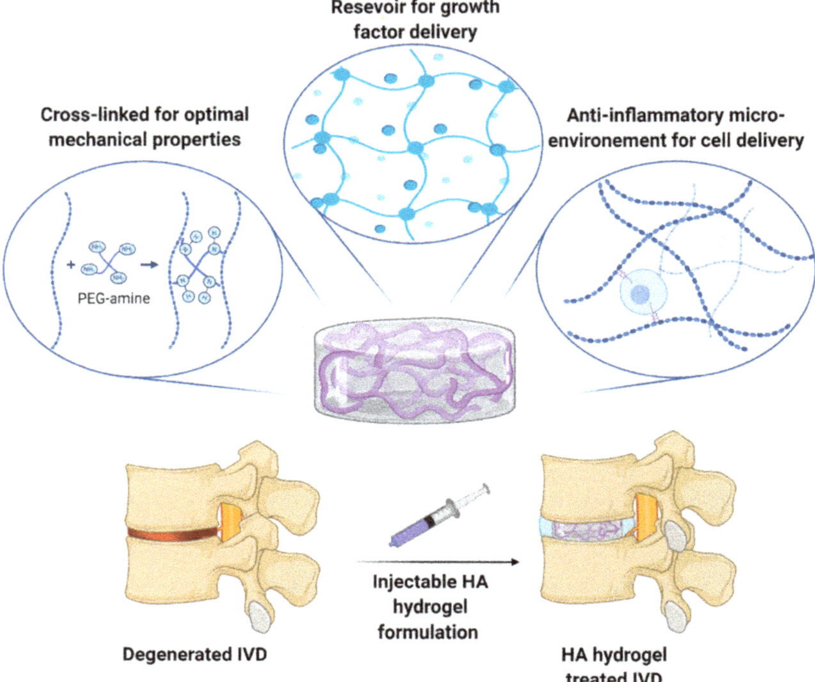

Figure 1. Different hyaluronic acid-based hydrogel platforms to treat the intervertebral disc degeneration. Hydrogel properties can be optimized to recapitulate the mechanical properties of the intervertebral disc (IVD), deliver therapeutics including growth factors and cells into the IVD. Hydrogels may be optimized to be injectable for minimally invasive treatment of the IVD.

Treatment of LBP in the early and moderate stages is mostly monitored through conservative treatments including bed rest, physiotherapy and exercise [20,21] which is followed by prescribing non-steroidal anti-inflammatory drugs (NSAID)s including Ibuprofen and COX-2 inhibitors (Celecoxib), that are effective in treating acute and chronic LBP comparing to placebo [22]. In addition to NSAIDs, muscle relaxants such as cyclobenzaprine are also used [23]. Moreover, opioids and benzodiazepines are prescribed to treat LBP. However, they have a short-term effect and need to be prescribed for no longer than four weeks because of the risk of developing dependency [24,25]. Furthermore, epidural and systemic corticosteroids are recommended with clinical trials exhibiting heterogeneous results, with the latest outcomes indicating that glucocorticoid receptor-specific steroids may be more effective comparing to mineralocorticoid-targeting steroids [26].

Although most of the LBP patients will recover after an acute phase, 10% may develop chronic low back pain [27]. In chronic stages, generally, surgery is required because the IVD is unable to cure itself [28]. Existing surgical interventions include discectomy and spinal fusion, or—occasionally in the cervical spine—a total disc replacement is indicated [29]. Patients who had discectomy show improvement when compared to nonsurgical therapy in short-term follow-ups; however, long term follow-ups (beyond two years) indicate no significant difference in the outcomes [30]. Spinal fusion is the typical surgical approach for chronic LBP. Although nonsurgical treatment for chronic LBP may be useful in a particular cohort, lumbar fusion remains more effective in alleviating pain and decreasing disability than commonly used nonsurgical treatments [31]. Disc replacement is also used to tackle the incidence of the disease in the adjacent segments due to the acute variation in motion segment mechanics linked with anterior cervical discectomy and fusion [32]. There is minimal evidence that suggests early surgical intervention improves long-term results in patients with lumbar disc herniation and radiculopathy that did not progress yet into neurologic deficit [30,33].

These treatments are not deemed curative, nor do they reverse the underlying pathology but rather aim to achieve symptomatic relief. Therefore, tissue-engineering approaches through developing biomaterial-based platforms to deliver active therapeutic products such as mesenchymal cells, inhibitory molecules and growth factors (GF) are being investigated to repair the degenerated IVD and revert it to a healthy state [34]. In the last few decades, hyaluronic acid (HA) was broadly used in treating osteoarthritis, popularised due to its anti-inflammatory, analgesic and extracellular matrix enhancing properties [35–37]. Hence, researchers shifted their interest towards treating the intervertebral disc using different HA compositions to deliver cells and therapeutics in vitro and in vivo (Figures 1 and 2. Therefore, the objective of this review article is to summarise the different HA-based platforms used to date in treating IVD degeneration and LBP. We also summarised the biomarkers and signalling pathways involved in disc degeneration. Moreover, we recapitulate the most popular preclinical models utilized so far in the field of IVD degeneration, emphasizing the fundamental need for more representative models to replicate human pathophysiology.

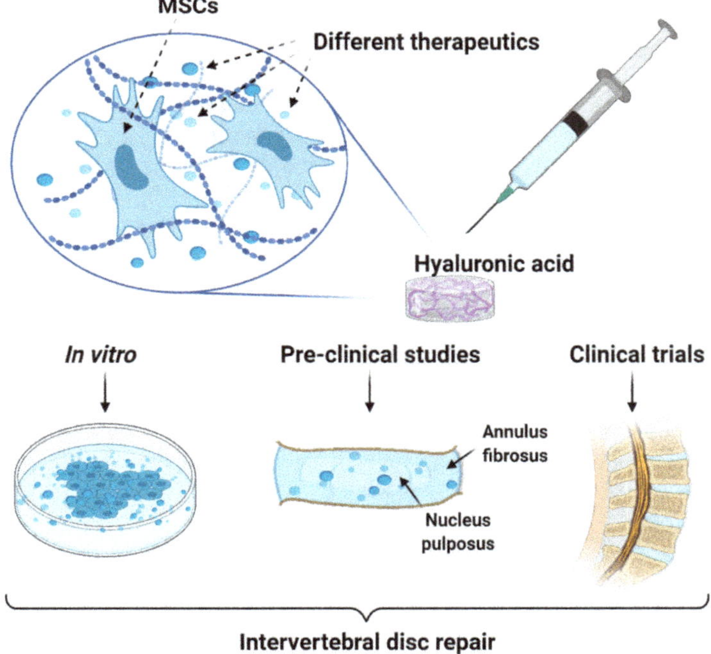

Figure 2. The role of hyaluronic acid as a platform for cell and drug delivery in the IVD. HA acts as a vehicle for adipose tissue and bone marrow isolated mesenchymal stem cells (MSC)s and drug delivery for disc repair. This schematic summarises the role of HA as a treatment for low back pain (LBP) and IVD degeneration in cell culture, preclinical studies and clinical trials. HA has been successful in disc repair through its anti-inflammatory, anti-apoptotic, analgesic and matrix modulatory characteristics.

2. Biomarkers and Signalling Pathways Associated with the Degenerated Intervertebral Disc

The literature associates IVD degeneration with dysregulated quantities of inflammatory cytokines produced by infiltrating macrophages, neutrophils, T cells, as well as disc cells [38]. During disc degeneration, the infiltration of mast cells and macrophages is enhanced by the vascularisation of the AF, which leads to amplified propagation of inflammatory signals, consequently, leading to LBP [39]. The inflammatory molecules secreted in the IVD, specifically prostaglandin E2 (PGE2), interferon-gamma (IFN-γ), tumour necrosis factor-alpha (TNF-α), and interleukin one alpha/beta (IL-1α/β), interleukin (IL)-17, IL-10, IL-6, IL-8, IL-2, IL-4, [38], stimulate the breakdown of the extracellular matrix (ECM), cell senescence, autophagy and apoptosis of the disc cells [38,40,41]. Pain sensation was also correlated with up-regulated nitric oxide and nitric oxide synthase [39]. Furthermore, the induced pro-inflammatory cytokines were found to stimulate the expression of neurogenic molecules, such as nerve growth factor (NGF) and brain-derived neurotrophic factor (BDNF), which may support the ingrowth of nerves as well as enhance nociception in dorsal root ganglia (DRG)s [38,41]. Hence, chronic inflammation prompts permanent structural and biochemical changes in the IVD comprising ECM degradation, vascularisation and innervation that ultimately leads to LBP [42]. Increased chemokine expression, especially C-C motif ligand 2 (CCL2), has been reported in the degenerated IVD [43]. Cultured AF cells from Thompson-graded discs of grade II-III were able to stimulate higher expression of CCL2 upon treatment with IL-1β than TNFα unlike AF cells from grades IV-V which showed a similar response to IL-1β and TNFα treatment in terms of CCL2 expression [44].

Different key dysregulated genes and proteins were discovered in the degenerated human AF such as interferon-stimulated genes (ISGs), which are interferon-induced proteins with tetratricopeptide repeats (IFIT)1, IFIT2, IFIT3 as well as Insulin-like growth factor-binding protein 3 (IGFBP3) [12]. Overexpression of the anti-proliferative IFIT3 protein and the pro-apoptotic IGFBP3 can negatively regulate the cell fate directly or indirectly, to induce growth arrest and apoptosis in AF cells [45–48]. IFIT3 can act as an anti-proliferative molecule through inducing the cyclin-dependent kinase inhibitors, such as cyclin-dependent kinase inhibitor 1A and cyclin-dependent kinase inhibitor 1B which interact with cyclin dependent kinases (CDK)s in the G1 phase and prevent the cell from proceeding into the S phase of the cell cycle [45]. IFNα was also able to induce IGFBP3 through the signal transducer, and activator of transcription 1 (STAT1) via cascades that have Insulin-like growth factor (IGF)-related anti-proliferative properties or IGFBP3 directly or indirectly induced pro-apoptotic consequences [46–48]. IGFBP3 is a vital controller of IGF availability that inhibits IGF activity, which is associated with osteoarthritis [49]. Moreover, IGFBP3 can sensitize cells to apoptosis indirectly through TNFα and IFNγ [50,51]. Several signalling pathways have also been identified in the disc as being aberrantly regulated, which are illustrated in Table 1.

3. Hyaluronic Acid in Treating the IVD: Structure, Synthesis and Turnover

HA, which was first isolated in the 1930s, is a distinctive glycosaminoglycan, which, unlike other glycosaminoglycans such as heparan sulphate, chondroitin sulphate, and keratan sulphate, is non-sulphated [52]. It is widely used in treating osteoarthritis [36,53–58] and in tissue-engineering approaches for disc repair [59,60]. HA (Figure 3) is an abundant polymer of the ECM with a molecular weight of approximately 10^6–10^7 Da in vivo [61]. HA molecule comprises of chains of constantly repeating disaccharides, glucuronic acid and *N*-acetylglucosamine, which are symbolized by GlcA-β (1$\rightarrow$3) and GlcNAc-β (1$\rightarrow$4) respectively (Figure 3) [62]. On the intracellular aspect of the cell membrane, synthesis of HA occurs through the hyaluronan synthases (HAS) family, comprised of three types of synthases, activated via different GFs to produce diverse forms of polymerized HA [63]. HAS1 and HAS2 produce the high molecular weight (HMW) of HA of 2–4 $\times$ 10^6 Da, while the low molecular weight of HA (LMW) of 1 $\times$ 10^5–1 $\times$ 10^6 Da is produced by HAS3 [63]. These enzymes are regulated by miRNAs [64] and GFs such as TGF-β [65]. Once synthesized, HA is secreted either on the cell surface from the plasma membrane or into the ECM because of its large size [63,66]. The body composition of

HA is 15 g [67]. Daily, 33% of HA is turned over through several roots: (i). via the lymphatic system or (ii). via the circulation through the liver or (iii). lysosomal digestion after HA binds to its receptor and is internalized [68]. Furthermore, HA turnover can take place as a result of pathological conditions such as tissue injury and remodelling as a result of oxidative stress and free radicals, HA internalization and degradation, and clearance of HA from tissues via the lymphatics and vascular system [69]. Regarding the receptors through which HA binds to be cleared by the liver and lymph nodes, there are particular receptors such as the hyaluronan receptor for endocytosis 1 (HARE1), lymphatic vessel endothelial hyaluronan receptor 1 (LYVE1) and CD44. From the previously mentioned three receptors, HARE1 is considered the major role player through the lymph and vasculature [68]. HA is internalized and degraded when it cannot be removed by the vasculature and lymph [69,70]. In this process, first HA binds to one of its receptors such as receptor for HA-mediated motility (RHAMM), layilin or CD44 after which it is degraded by hyaluronidase-2 [65,69,70], then it is digested into fragments through *N*-glucuronidases and *N*-acetylglucosaminidases to completely break down the remaining glycans into monosaccharides [70].

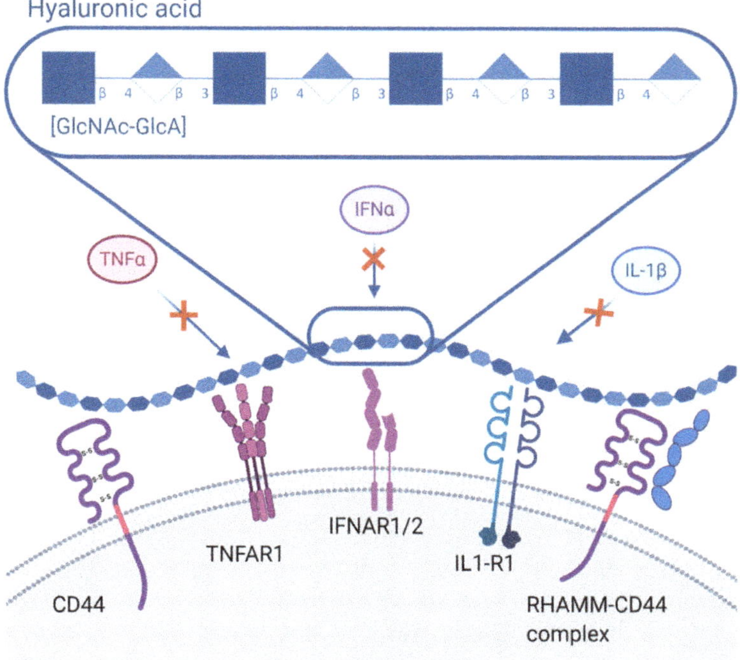

Figure 3. The anti-inflammatory role of hyaluronic acid (HA). HA molecule (on the top) is represented according to its international symbols by repeated units of glucuronic acid (GlcA-β1-3) and N-acetylglucosamine (GlcNAc-β1-4) disaccharides. The anti-inflammatory role of HA is mediated through reacting with its several receptors such as CD44 and receptor for HA-mediated motility (RHAMM) preventing the cascades signalled by inflammatory cytokines such as IFNα, TNFα and IL1β through deterring the contact with their receptors.

The Key Function of Hyaluronic Acid in Repairing the Intervertebral Disc

The biochemical properties of the HA molecule are entirely dependent on its size [69,71,72]. Within tissues, the HA molecule interacts with various proteins comprising of hyaladherins such as CD44, RHAMM (CD168) (Figure 3). In contrast, inside the ECM, it interacts with N-terminals of proteoglycans such as aggrecan (ACAN), versican, brevican, neurocan and TNFα-stimulated gene-6

(TSG-6) protein [69,73]. These fundamental interactions are essential for HA function in terms of cell-to-cell communication and signalling [65,74]. The mechanical properties of HA can be attributed to its high molecular weight structure. HA is capable of absorbing water approximately 10–10^4 times its mass as it is negatively charged and is considered an osmotically active molecule [63,64,75,76]. Due to its large volume while hydrated, it can fill large spaces, therefore absorbing shocks and acting like a lubricant [75,77].

HA-CD44 interactions mediate endocytic activity and activate Rho and Rac1 GTPases to regulate cytoskeletal organization [78]. CD44 activation also activates src-related tyrosine kinases to induce cell proliferation through NF-κB signalling [79]. RHAMM activates pp60c-src tyrosine kinase to modulate focal adhesions for RHAMM-mediated cell mobility [80]. Its ability to form large extracellular networks through cell–matrix binding inhibits cytokine binding to prevent downstream signalling (Figure 3) [75,81]. Therefore, HMW HA is capable of inhibiting the signalling of inflammatory cytokines and matrix-degrading enzymes (TNF-α, IL-8, iNOS, aggrecanase 2, matrix metalloproteinases (MMP)s) that were evident upon treating chondrocytes with HMW HA, which also resulted in inhibition of phagocytosis and macrophage activation [82,83]. In tissue injury and remodelling, enzymes break down HMW HA resulting in fragments of LMW HA which can induce phosphorylation of p38 or p42 or p44 mitogen-activated protein kinase (MAPK) and NF-κB which are associated with toll-like receptor (TLR)4 signalling [63,84]. Depending on the tissue type and the size of LMW HA fragments (f-HA)s, pro-inflammatory cytokines can be induced [63,81]; therefore, angiogenesis and tissue remodelling is stimulated [69,71]. It was discovered that human disc cells treated with 6–12 disaccharide fragments of HA can dramatically up-regulate catabolic enzymes such as IL-1β, IL-8, IL-6, MMP1, MMP13, and cyclooxygenase (COX2). Furthermore, it was identified that TLR-2 mediated the role of f-HAs in stimulating the protein expression of IL-6, and both IL-6 and MMP-3 were supported by the MAPK signalling pathway [85].

4. Translational Preclinical Models Used to Investigate the Intervertebral Disc Degeneration

Studying IVD degeneration using human subjects is challenging because of the ethical restrictions hindering the access of the clinical data as well as differentiating the different factors influencing the disc degeneration which include ageing, genetic predisposition, mechanical loading, and inflammation [86]. Therefore, developing reliable and representative animal models is necessary. As a result, different animal models emerged comprising rodents, canines, porcine, rabbit, sheep, and goat even though these animals are different in their anatomy, development, and tissue mechanical characteristics compared to human discs [87,88]. Such models were utilized to evaluate different hyaluronan-based therapeutic approaches for the IVD repair (Table 2). Several disc degeneration replicas are present to study disc degeneration in in vivo models: spontaneous (genetic aberration induces early disease onset) [89] puncture (physical trauma activating inflammatory response) [90,91] mechanical loading (non-traumatic induction) [92,93] and biochemical (Chondroitinase ABC treatment) [94]. Preclinical models were also used to introduce novel surgical approaches, especially in large animals [95]. Rodent tails have been extensively used to study the pathophysiology of the IVD degeneration and assess the effect of different therapeutics because they are easily accessible with minimal interference with the surrounding tissues and their normal physiological functions [92,93,96–104]. The typical age of the rats used in such studies is three-months-old and above because they can only reach skeletal maturity at that age [105]. Notochordal cells can be found in the IVD of rats throughout adulthood while they are lost in humans [93]; however, these cells are lost in many degenerative rat-tail models due to factors such as mechanical loading [106]. Yurube et al. developed a model of IVD degeneration in a rat-tail in vivo model by using Ilizarov apparatus to load the tail discs, which was first used by Iatridis et al. in 1999 [107]. The Ilizarov-type apparatus was set on the coccygeal 8–9 (C8/9) discs to induce loading on C9/10 while C10/11 remained unloaded as a control. As a result, it was found that after seven day, the notochordal cells were lost in the rat-tail discs upon static compression up to day 56. Moreover, on day seven, it was noted that the extrinsic apoptotic pathway was activated in the AF and NP that progressively decreased. In contrast, the intrinsic mitochondrial pathway remained continuously active until day 56 [108]. Under disc compression, ECM turnover was explained through tissue inhibitors of metalloproteinase one (TIMP-1), MMP-2 and MMP-13 up-regulation [109,110]. In parallel, it was reported that disc loading leads to imbalances in the manifestation of MMPs, a disintegrin and metalloproteinase with thrombospondin motifs (ADAMTS), and TIMPs in rat-tail discs [92]. Small and large animal models were utilized over the decades to target dysregulated biomarkers and signalling pathways (Table 1) to mitigate the IVD degeneration using HA-based biomaterials summarised in Table 2. Different HMW HA and HA-based therapeutic platforms were tested in vitro, in vivo and human trials to repair the disc (Figure 2).

Table 1. Dysregulated signalling cascades in the degenerated IVD.

Signalling Cascades	IVD Segment	Reference
Transforming growth factor-beta (TGF-β)	IVD	[111]
Mammalian target of rapamycin (mTOR)	IVD	[112]
Wingless-related integration site (Wnt)	AF, NP, end Plate (EP), growth plate (GP)	[113–118]
Nuclear factor-kappa beta (NF-κB)	AF, NP, IVD	[119–125]
Nerve growth factor (NGF)	AF, NP	[126–131]
Mitogen-activated protein kinase (MAPK)	AF, NP	[132–138]
Notch signalling (NOTCH)	AF, NP	[139]
Fibroblast growth factor (FGF)	AF, NP	[140,141]
Interferon-alpha signalling canonical pathway (IFNα)	AF	[12,60]

Table 2. Different hyaluronic acid compositions evaluated in preclinical and clinical trials for IVD repair.

HA Composition and Cross-Linking Method	Species/Platform	Objectives	Outcome	Reference
Collagen-I and HA Ultraviolet (UV) cross-linking	Bovine In vitro	To assess the efficacy of collagen-I and HA injection in IVD repair, both including and excluding GF.	• In terms of ECM components, there were no evident changes between AF and NP in synthesizing collagen, aggrecan and leucine-rich proteoglycans. • In both types of cells, TGF-β1 induced ECM synthesis. • According to the outcomes, the effect of TGF-β1 and primary fibroblast growth factor (bFGF) combination was more effective than that of TGF-β1.	[142]
HA and chondroitin sulphate (CS) hydrogel Cinnamic acid + UV cross-linking	Rabbit In vivo	To evaluate the cross-linked HA and CS's effect on the disc regeneration.	• Histological staining showed that the inner AF of the discs that were treated with cross-linked HA or CS have thick Safranin-O staining which is than those injected with 1% sodium hyaluronate or phosphate buffer saline (PBS). • According to the magnetic resonance imaging (MRI) results, the injection of cross-linked HA and CS can neutralize the intensities of MRI signals detected by injured discs.	[143]
Hylan G-F20 (Synvisc) Avian derived HA Cross-linked by divinyl sulfone	Primate In vivo	To evaluate if HA can affect the degenerative pathway in non-human primates after undergoing nucleotomy.	• Comparing to the control discs, Hylan G-F20 administered segment height was reduced of around 10% only unlike the ones undergone nucleotomy. • Comparing to the control, standard nucleotomy and Hylan B treated functional spinal unit resulted in significant disc space narrowing. • According to the MRIs, CT scans, and macroscopic evaluation, Hylan G-F20 administration suggested to be significantly effective in comparison to the standard nucleotomy when compared to the control. • Overall, Hylan G-F20 can be a postoperative therapeutic approach to halt disc degeneration after nucleotomy.	[144]
HMW HA PEG-amine cross-linked with (1-ethyl-3-(3-dimethylaminopropyl) carbodiimide hydrochloride) (EDC)/N-hydroxysuccinimide (NHS)	Bovine Ex vivo	To identify HA's matrix modulatory and anti-inflammatory properties.	• The outcomes revealed that HMW HA demonstrated anti-inflammatory properties by decreasing the expression of IGFBP3 and IFIT3 as well as the downstream signalling molecules of IFN$\alpha_2\beta$. • HMW HA was also able to up-regulate the ECM aggrecan and collagen-1.	[145]

Table 2. *Cont.*

HA Composition and Cross-Linking Method	Species/Platform	Objectives	Outcome	Reference
HMW HA **PEG-amine cross-linked with EDC/NHS**	Bovine In vitro	To identify the effect of HA in an IL-1β inflammation model of NP cells.	• HA hydrogel was stable in PBS and retained more than 40% mass under degradation by enzymes. • There was no observed cytotoxic effect of HA hydrogel on NP cells for the duration of the culture up to 7 days. • Gene expression analysis revealed that upon treating cells with cross-linked HA, there was a significant down-regulation of IL-1R1, myeloid differentiation factor 88 (MyD88), NGF and BDNF. • Upon treatment with HA, CD44 receptors were up-regulated in the NP cells after which suggested that HA can have an anti-inflammatory effect by reacting with CD44 on the cells.	[59]
Glycidyl methacrylate (GM)-HA **HA with collagen devoid matrix** **UV cross-linking**	Porcine In vivo	To evaluate HA/collagen hydrogel in a porcine nucleotomy model.	• IVD degeneration was induced after nucleotomy using a 16G needle. • An inflammatory reaction to the material was detected. • Inflammation leads to annular scarring. However, the hydrogel was successful in deterring re-herniation. • HA-treated discs increased expression of collagen-I collagen-II, MMP13 and TIMP1.	[146]
Hyaluronan gel (Durolane®)	Porcine In vivo	To evaluate HA as a vehicle for different types of cells used for IVD repair.	• The hyaluronan gel promoted cell proliferation in vitro. • Synthesis of collagen-II was observed in the MSCs and chondrocytes which survived in the porcine IVDs up to six months. • MRI scanning revealed significant changes in the endplate showing severe IVD degeneration in animals after six months of transplanting various types of implants. • Bone mineralization was evident by positive staining. • In vivo tested hyaluronan gel was not successful as a cell carrier.	[147]
HMW HA **PEG-amine cross-linked with EDC/NHS**	Rat In vivo	To assess HA's effect on the IFNα expression as well as the extracellular matrix modulation by conducting proteomic data analysis.	• IFNα was significantly down-regulated on days 7, 28 and 56 in the discs bearing injury, which were implanted with HMW HA. • Caspase 3 was found down-regulated on days 7, 28 and 56 in the discs bearing injury which were implanted with HMW HA. • In ECM, two key ECM proteins were found up-regulated, which are aggrecan and hyaluronan and proteoglycan link protein 1 (HAPLN1) over the different time points in response to HMW HA treatment. • Regarding the glycosylation pattern, we found sialylation, which is an indicator of inflammation was down-regulated on days 7, 28 and 56 when the injured discs were treated with HMW HA.	[60]

Table 2. *Cont.*

HA Composition and Cross-Linking Method	Species/Platform	Objectives	Outcome	Reference
HMW HA **PEG-amine cross-linked with EDC/NHS**	Rat In vivo	To assess HA's effect as an analgesic and anti-inflammatory therapeutic in a pain model of IVD.	• HA reduced nociceptive behaviour and inhibited, hyper-innervation. • HA altered the glycosylation pattern of the degenerated disc. • HA modulated key inflammatory signalling pathways; therefore, attenuating inflammation and its effect on the extracellular matrix.	[148]
Aminated hyaluronic acid-g-poly N-isopropyl acrylamide-(AHA-g-PNIPAAm)-gefitinib **EDC cross-linking**	Mice/Rat/ human In vivo	To assess the effect of AHA-g-PNIPAAm-gefitinib in mitigating IVD degeneration by intervening with the function of the epidermal growth factor receptor (EGFR).	• AHA-g-PNIPAAm-gefitinib suppressed EGFR activity, increased autophagy and ECM production, decreased MMP13 and prevented the progression of IVD degeneration in humans.	[149]
Hyaluronic and fibrin acid hydrogel (RegenoGel)–Fibroblast growth factor-18 **Fibrin cross-linking**	Human/ Bovine In vitro	To assess the effect of both FGF-18 and the FBG-HA hydrogel on NP regeneration.	• FBG-HA-treated human NP cell cultures showed an up-regulation in the expression of carbonic anhydrase XII and collagen-II after 7 and 14 days, respectively. • Increased glycosaminoglycan (GAG) release was noted over 14 days in the conditioned medium. • From day 7 to day 14, increased expression of ACAN was noted in bovine NP cells. • In human NP cells, FGF-18 up-regulated CA12. • According to the histology results: 1. Unlike human NP, bovine cells showed an increase in proteoglycan synthesis as a result of FGF-18. 2. FBG-HA hydrogel showed a significant effect on mitigating the degeneration. • In terms of cell proliferation and GAG synthesis, FGF-18 did not show any significant effect in the NP cells.	[150]
HYADD®4-G **hexadecyl amine of 5 × 10^2 to 7.3 × 10 kDa HA aliphatic amines bound to glucuronic acid at 2% substitution**	Rabbit In vivo	To evaluate the effect of HYADD®4-G in IVD repair.	• Compared to the control, saline injections increased the disc height by 50% of the initial disc height while HYADD®4-G administration increased it by over 75%. • MRIs showed that HYADD®4-G administration leads to significantly higher water absorption compared to the control treatment. • HYADD®4-G injected discs (83.3%) were restored to grade I based on the MRI grading system (Pfirrmann). • In terms of cellularity, tissue organization and comparing to saline injection, HYADD®4-G administration showed significantly decreased scores of IVD degeneration. • No inflammatory reactions were observed to HYADD®4-G injection into the discs.	[151]

Table 2. *Cont.*

HA Composition and Cross-Linking Method	Species/Platform	Objectives	Outcome	Reference
Platelet-rich plasma (PRP) and HA Batroxobin (BTX) cross-linking	Human In vitro	To assess whether PRP/HA/BTX blend is the best delivery vehicle for MSCs.	• There was higher MSCs viability and proliferation in the hydrogel. • Compared to the control, significantly higher GAG production was observed in MSCs in culture, including or excluding TGF-β1. • Histology and gene expression analysis revealed that MSCs within the hydrogel treated with TGF-β1, differentiated into chondrocyte-like cells expressing ACAN, COL2 and SOX9.	[152]
Hyaluronan oligosaccharides (HA-oligos)	Ovine In vitro In vivo	To evaluate the effect of HA-oligos in inducing ECM anabolic gene expression and metalloproteinase.	• AF cells cultured in 2D respond mildly to the HA-oligo, where proMMP-2 levels were slightly up-regulated while MMP-9 was not changed; in NP cells, ProMMP-2 increased in a dose-dependent manner. • In AF alginate bead culture, the active form of MMP-9 and pro-MMP-2 was up-regulated until day 10; while in NP alginate bead culture, active MMP-9 was up-regulated on day ten while proMMP-2 continuously changed into the active MMP-2 in days 7–10. • In the non-stimulated NP cultures, MMP-2 and MMP-9 activity were down-regulated over 2–10 days. • HA-oligo was shown not to be cytotoxic because of the high disc cell viability. • RT-PCR showed that MMP1, MMP13 and ADAMTS1 as well as the matrix genes COL1A1 and COL2A1 and ACAN were up-regulated in the NP via HA-oligos; while in the AF ADAMTS1, ADAMTS4, ADAMTS5, MMP13, and COL2A1 and ACAN expression were decreased. • Histological analysis showed that in the outer lesion zone, collagen-I enhanced remodelling by the HA-oligo treatment.	[153]
Fibrin/HA (FB/HA) Transglutaminase cross-linking	Goat In vivo	To evaluate the safety and the effect of injecting bone morphogenic protein 2 (BMP-2) and BMP-2/7 incorporated into (FB/HA) intradiscally for the purpose of disc repair in a mild disc degeneration model in the goat.	• No adverse effects or heterotopic bone formation were observed. • After induction of mild disc degeneration, a significant disc height decrease was observed in the radiographs. • MRI T2* results were correlated with histology and biochemistry outcomes. • Intervention groups did not show any significant changes. • Although BMP-2 and BMP-2/7 turned to be safe, there was no evidence of disc regeneration.	[154]

Table 2. *Cont.*

HA Composition and Cross-Linking Method	Species/Platform	Objectives	Outcome	Reference
Gelatin and HA methacrylate (GelHA) UV cross-linking	Rat In vitro In vivo	To evaluate the effect of GelHA on enhancing the adipose stromal cells (ASCs) differentiation.	• GelHA hydrogel induced ASCs differentiation which was detected through up-regulation of the NP markers in the group of GelHA and ASCs compared with the control and ASC alone. • Compared with normal cultured cells, the group including GelHA and ASCs up-regulated TGF-β1 and TGF-β RII genes in 14 and 21-days. • neutralizing antibody suppressed the expression of NP matrix proteins in ASCs in vitro. • In vivo trials in rats showed that the hydrogel composed of GelHA and ASCs enhanced IVD repair through up-regulating NP matrix synthesis and significantly increasing the disc height.	[155]
Collagen and HA incorporated fibrin-based hydrogels **Fibrin-thrombin cross-linking**	Porcine In vitro	To evaluate the effect of the HA and collagen incorporated fibrin hydrogel in NP-like matrix synthesis.	• High fibrin concentrations enhanced cell viability and proliferation. • In Fibrin–collagen hydrogels collagen synthesis was also detected. • HA enhanced the chondrocyte proliferation and induced the proteoglycan synthesis in the ECM. • Incorporation of HA enhanced GAG synthesis while led to the suppression of total collagen development at higher concentrations. • 5 mg/mL HA was the most optimal concentration of HA for disc regeneration, and this is because it could support NP-like matrix synthesis of the articular chondrocytes.	[156]
Combined adipose-derived mesenchymal stem cells (AT-MSCs) and HA derivative **Butanediol diglycidyl ether cross-linking**	Human Clinical Trial	To assess the effect and tolerability of administration of AT-MSCs combined with HA derivative intradiscally in chronic discogenic LBP patients.	• The clinical trial did not result in any procedural or stem cell-related adverse changes during the follow-ups of the 1st year. • In terms of the Short Form 36 (SF-36), visual analogue scale (VAS), and the Oswestry disability index (ODI) scoring, no significant changes were observed between the low and high cell dose groups. • Three patients out of the six with the cases 4, 8, and 9 who had significant improvement which was prevalent through SF-36, VAS, ODI and scoring, were identified having higher water content in MRI scans. • Results showed that it is safe to implant HA with AT-MSCs in patients with chronic LBP.	[157]
Sodium hyaluronate (SH)	Human Clinical Trial	To assess the effect and the tolerability of SH administration in comparison with glucocorticoids triamcinolone acetonide (TA) in treating non-radicular back pain.	• After the administration of SH and throughout the follow-up period, there were no unfavourable changes. • The SH treatment, the effect of which was equal to a course of TA injections, improved significantly patients' quality of life of with non-radicular pain via reducing pain and improving function. • Comparing to TA-treated, SH-treated patients expressed more extended-term benefits.	[158]

There is an increasing requirement for three-dimensional (3D) scaffolds to regenerate IVDs. HA has a diverse range of applications when integrated into such 3D structures. Different materials were incorporated with HA to form composites with biocompatible properties including biological performance, stiffness, and degradation both in vitro and in vivo. Such composites, including collagen, Fibrin and chondroitin sulphate [142,143,146,156], have been investigated in order to tackle extracellular matrix degradation by enhancing the synthesis of key matrix modulating proteins, GAGs and increasing cellular viability. Pre-clinical models of IVD degeneration to test the efficacy of HA-based materials have been developed in mice [155], rats [60], rabbit [143], sheep [153], goats [154], pigs [156], cows [142], and primates [144]. HA formulations have been cross-linked using many methods including UV cross-linking [142], divinyl sulfone [144], polyethyleneglycol (PEG)-amine with EDC (1-ethyl-3-(3-dimethylaminopropyl)carbodiimide), NHS (N-hydroxysuccinimide) cross-linking [59], fibrin [150], batroxobin [152], butanediol diglycidyl ether [157] and enzymatic cross-linking using transglutaminase in a fibrin/HA material [154]. HA hydrogels utilized in treating the different parts of the IVD, including AF and NP, not only acted as composite materials but also as a reservoir for growth factors and therapeutic drugs [142,150,154], as well as a cell delivery vehicle [147,152,155].

In summary, these studies report on the hydrophilic properties of HA to restore disc height [151] and MRI signal intensity [144], the ability of HA to promote ECM synthesis [147] in resident or injected cells, and of course the anti-inflammatory [145] and anti-nociceptive [148] properties of HA in the setting of IVD degeneration. Several clinical trials investigating the efficacy of HA-based materials are ongoing, with few reporting results at this early stage. However, initial results have been promising, where MSCs delivered in a HA carrier system demonstrated an excellent safety profile with several patients reporting improved VAS and ODI reflecting better MRI signal intensity [157].

Furthermore, HA is being trialled in extra-discal injections in cases of facet joint degeneration [158]. HA has demonstrated equivalent efficacy to glucocorticoid administration in this patient cohort. Many studies are yet to report results, including a bone marrow-derived MSC (BM-MSC)/HA system in a Phase 2 trial (NCT01290367), Rexlemestrocel-L combined with hyaluronic acid in Phase 3 (NCT02412735) and discogenic cells in a HA delivery system with study arms in USA and Japan in Phase 1 (NCT03347708). There is much excitement surrounding these systems, given the proven HA preclinical efficacy.

5. Discussion and Future Directions

Discogenic back pain is a prevalent disease described by the degeneration of one or more of the intervertebral discs causing pain, dramatically affecting the quality of life. LBP generation is initiated by low numbers of native IVD cells, which leads to extracellular matrix disintegration, and vascularisation promoting inflammation and low back pain sensation. It was found that pro-inflammatory cytokines such as IFNα induce cellular arrest and apoptosis of the disc cells through downstream targets such as IFIT3 and IGFBP3; therefore, leading to low cellularity of the disc [12]. Moreover, the up-regulation of different inflammatory cytokines can induce the disintegration of the ECM, synthesis of chemokine and, ultimately, changes in IVD cells [38].

For the ideal HA-based therapeutic to be identified, it needs to be compatible and supportive to the disc microenvironment in general and target inflammation and downstream signalling cascades that lead to innervation, vascularisation and pain sensation in particular. In the last few decades, HA and a combination of HA-based biomaterials were tested in preclinical disc degeneration models and human clinical trials. Multiple cross-linking techniques have been employed, including UV cross-linking [142], chemical cross-linking [145] and combination devices that also utilize molecular self-assembly [146]. There is no apparent consensus on optimal cross-linking methods, nor has there been detailed investigation into the retained bioactivity using these cross-linking methods. Although it could be challenging because of the complexity of the disease, research has focused on a different combination of materials that could be quite promising for IVD degeneration. HA, a sizeable hydrophilic molecule that can absorb shocks, is a fundamental component of ECM which developed to be a promising platform for next-generation therapeutics. Through the use of various combinations of

HMW HA and processing techniques, inflammation, innervation, as well as pain, can be reduced in the disc [59,60,148]. The material design process must further consider the degradation products of these materials and ensure these metabolites do not induce pro-inflammatory signalling in long-term follow-up. While LMW HA fragments induce inflammation, HA can be modified to resist degradation and mitigate the effects of fragmentation [84,159].

In this review, we identified a series of studies focused on assessing the effect of different compositions of HA as a therapy for aberrantly expressed enzymes, proteins or signalling pathways identified in the degenerated IVD. It was found that HA was effective in lowering the IFNα inflammatory signalling pathway and downstream key apoptosis and cell senescence regulators such as IFIT3 and IGFBP3 in the AF [60]. Similarly, HA was found to be capable of lowering the NF-κB signalling pathway and its downstream targets NGF and BDNF in the NP [59].

In addition to its use as anti-inflammatory, anti-apoptotic and analgesic biomaterial in the disc, HA was also used as a carrier for MSCs in vivo [147,155]. The results of recent clinical trials [157] supported that the implantation of adipose-derived MSCs embedded in HA derivative (Figure 2) is non-toxic and endurable in LBP and could enhance the patients' quality of life by reducing the back pain and enhancing their function [158]. HA encapsulation of MSCs injected into the disc offers structural integrity and allows for cytoskeletal attachment through CD44 binding [59,160]. HA may also block noxious signalling pathways present in the degenerated IVD to preserve cell phenotype. Ongoing clinical trials are using HA as a delivery medium for BM-MSCs and discogenic cells to improve cell survival and reduce inflammation [161].

Furthermore, HA has been used as a vehicle to deliver different types of therapeutics (Figure 2) such as TGF-β1 [142], BMP-2/7 [154] and gefitinib [149]. The swelling capabilities and tuneable mechanical characteristics of HA hydrogels popularise them for the tissue-engineering approach for direct repair of the disc or as a vehicle for cell therapy of small-molecule drug delivery. HA also acts as a vehicle for GF regulating protein synthesis and cell proliferation [162]. HA and its derivatives effectively bind GF through electrostatic interactions to induce signalling and modulate ligand–receptor activity [162]. The degradation of the HA macromolecule by native tissue hyaluronidases can enhance small-molecule release [163].

In general, the majority of the HA applications were investigated in small or large animal models which are limited in their translation of IVD degeneration in humans [93]. These models were not spontaneous disc degeneration models but induced acute inflammatory models through puncture [164,165] or enzymatic digestion [166]. Therefore, these models might not be ideal for testing therapeutics because they induce rapid degeneration of the disc, rather than a natural onset of disease. Hence, a model of IVD degeneration which replicates the human IVD degeneration pathology is necessary.

Next, what is expected in the prospective studies is the assessment of the effect of potentially identified HA-based therapeutics on a larger scale and in the context of the cross-talk of multiple inflammatory signalling pathways to address the complex environment of the degenerative disc disease and support achieving the ultimate goal which is finding a universal HA-based treatment for intervertebral disc degeneration. HMW HA which was widely used by numerous studies as an anti-inflammatory and matrix enhancing molecule [167,168] must be further investigated in a spontaneous model of disc degeneration in a large animal model to address the anti-inflammatory effect on combined key identified inflammatory signalling pathways such as TNFα, IFNα and IL1-β as illustrated in Figure 3. Another critical factor to focus on is the development of a robust minimally invasive technique for HA administration without requiring open surgery. Although there are challenging obstacles ahead, HA-based injectable hydrogels are strong candidates as effective treatments for disc degeneration.

31. Fritzell, P.; Hagg, O.; Wessberg, P.; Nordwall, A.; Swedish Lumbar Spine Study Group. 2001 Volvo Award Winner in Clinical Studies: Lumbar fusion versus nonsurgical treatment for chronic low back pain: A multicenter randomized controlled trial from the Swedish Lumbar Spine Study Group. *Spine* **2001**, *26*, 2521–2532. [CrossRef] [PubMed]
32. Guyer, R.D.; Ohnmeiss, D.D. Intervertebral disc prostheses. *Spine* **2003**, *28* (Suppl. 15), S15–S23. [CrossRef] [PubMed]
33. Weinstein, J.N.; Tosteson, T.D.; Lurie, J.D.; Tosteson, A.N.; Hanscom, B.; Skinner, J.S.; Abdu, W.A.; Hilibrand, A.S.; Boden, S.D.; Deyo, R.A. Surgical vs nonoperative treatment for lumbar disk herniation: The Spine Patient Outcomes Research Trial (SPORT): A randomized trial. *JAMA* **2006**, *296*, 2441–2450. [CrossRef] [PubMed]
34. Stergar, J.; Gradisnik, L.; Velnar, T.; Maver, U. Intervertebral disc tissue engineering: A brief review. *Bosn J. Basic Med. Sci.* **2019**, *19*, 130–137. [CrossRef] [PubMed]
35. Tashiro, T.; Seino, S.; Sato, T.; Matsuoka, R.; Masuda, Y.; Fukui, N. Oral administration of polymer hyaluronic acid alleviates symptoms of knee osteoarthritis: A double-blind, placebo-controlled study over a 12-month period. *Sci. World J.* **2012**, *2012*, 167928. [CrossRef] [PubMed]
36. Bowman, S.; Awad, M.E.; Hamrick, M.W.; Hunter, M.; Fulzele, S. Recent advances in hyaluronic acid based therapy for osteoarthritis. *Clin. Transl. Med.* **2018**, *7*, 6. [CrossRef] [PubMed]
37. Altman, R.; Bedi, A.; Manjoo, A.; Niazi, F.; Shaw, P.; Mease, P. Anti-inflammatory effects of intra-articular hyaluronic acid: A systematic review. *Cartilage* **2019**, *10*, 43–52. [CrossRef] [PubMed]
38. Risbud, M.V.; Shapiro, I.M. Role of cytokines in intervertebral disc degeneration: Pain and disc content. *Nat. Rev. Rheumatol.* **2014**, *10*, 44–56. [CrossRef] [PubMed]
39. Kadow, T.; Sowa, G.; Vo, N.; Kang, J.D. Molecular basis of intervertebral disc degeneration and herniations: What are the important translational questions? *Clin. Orthop. Relat. Res.* **2015**, *473*, 1903–1912. [CrossRef] [PubMed]
40. Quan, M.; Hong, M.W.; Ko, M.S.; Kim, Y.Y. Relationships between disc degeneration and autophagy expression in human nucleus pulposus. *Orthop. Surg.* **2020**, *12*, 312–320. [CrossRef] [PubMed]
41. Sampara, P.; Banala, R.R.; Vemuri, S.K.; Av, G.R.; Gpv, S. Understanding the molecular biology of intervertebral disc degeneration and potential gene therapy strategies for regeneration: A review. *Gene Ther.* **2018**, *25*, 67–82. [CrossRef] [PubMed]
42. De Geer, C.M. Cytokine involvement in biological inflammation related to degenerative disorders of the intervertebral disk: A narrative review. *J. Chiropr. Med.* **2018**, *17*, 54–62. [CrossRef] [PubMed]
43. Gschwandtner, M.; Derler, R.; Midwood, K.S. More than just attractive: How CCL2 influences myeloid cell behavior beyond chemotaxis. *Immunol. Front. Immunol.* **2019**, *10*, 2759. [CrossRef] [PubMed]
44. Gruber, H.E.; Hoelscher, G.L.; Ingram, J.A.; Bethea, S.; Cox, M.; Hanley, E.N., Jr. Pro-inflammatoryy cytokines modulate the chemokine CCL2 (MCP-1) in human annulus cells in vitro: CCL2 expression and production. *Exp. Mol. Pathol.* **2015**, *98*, 102–105. [CrossRef] [PubMed]
45. Xiao, S.; Li, D.; Zhu, H.Q.; Song, M.G.; Pan, X.R.; Jia, P.M.; Peng, L.L.; Dou, A.X.; Chen, G.Q.; Chen, S.J.; et al. RIG-G as a key mediator of theanti-proliferative activity of interferon-related pathways through enhancing p21 and p27 proteins. *Proc. Natl. Acad. Sci. USA* **2006**, *103*, 16448–16453. [CrossRef] [PubMed]
46. Foser, S.; Redwanz, I.; Ebeling, M.; Heizmann, C.W.; Certa, U. Interferon-alpha and transforming growth factor-beta co-induce growth inhibition of human tumor cells. *Cell Mol. Life Sci.* **2006**, *63*, 2387–2396. [CrossRef] [PubMed]
47. Kiepe, D.; Ulinski, T.; Powell, D.R.; Durham, S.K.; Mehls, O.; Tonshoff, B. Differential effects of insulin-like growth factor binding proteins-1, -2, -3, and -6 on cultured growth plate chondrocytes. *Kidney Int.* **2002**, *62*, 1591–1600. [CrossRef] [PubMed]
48. Del Monte, P.; Laurino, C.; Arvigo, M.; Palermo, C.; Minuto, F.; Barreca, A. Effects of alpha-interferon on insulin-like growth factor-I, insulin-like growth factor-II and insulin-like growth factor binding protein-3 secretion by a human lung cancer cell line in vitro. *J. Endocrinol. Investig.* **2005**, *28*, 432–439. [CrossRef] [PubMed]
49. Evans, D.S.; Cailotto, F.; Parimi, N.; Valdes, A.M.; Castano-Betancourt, M.C.; Liu, Y.; Kaplan, R.C.; Bidlingmaier, M.; Vasan, R.S.; Teumer, A.; et al. Genome-wide association and functional studies identify a role for IGFBP3 in hip osteoarthritis. *Ann. Rheum. Dis.* **2014**, *74*, 1861–1867. [CrossRef] [PubMed]

50. Baxter, R.C. Signalling pathways involved inanti-proliferativee effects of IGFBP-3: A review. *Mol. Pathol.* **2001**, *54*, 145–148. [CrossRef] [PubMed]
51. Fang, P.; Hwa, V.; Little, B.M.; Rosenfeld, R.G. IGFBP-3 sensitizes prostate cancer cells to interferon-gamma-induced apoptosis. *Growth Horm IGF Res.* **2008**, *18*, 38–46. [CrossRef] [PubMed]
52. Liang, J.; Jiang, D.; Noble, P.W. Hyaluronan as a therapeutic target in human diseases. *Adv. Drug Deliv. Rev.* **2016**, *97*, 186–203. [CrossRef] [PubMed]
53. Lai, H.Y.; Chen, Y.C.; Chen, T.J.; Chou, L.F.; Chen, L.K.; Hwang, S.J. Intra-articular hyaluronic acid for treatment of osteoarthritis: A nationwide study among the older population of Taiwan. *BMC Health Serv. Res.* **2008**, *8*, 24. [CrossRef] [PubMed]
54. Chou, C.L.; Li, H.W.; Lee, S.H.; Tsai, K.L.; Ling, H.Y. Effect of intra-articular injection of hyaluronic acid in rheumatoid arthritis patients with knee osteoarthritis. *J. Chin. Med. Assoc.* **2008**, *71*, 411–415. [CrossRef]
55. Migliore, A.; Granata, M. Intra-articular use of hyaluronic acid in the treatment of osteoarthritis. *Clin. Interv. Aging* **2008**, *3*, 365–369. [CrossRef] [PubMed]
56. Chircov, C.; Grumezescu, A.M.; Bejenaru, L.E. Hyaluronic acid-based scaffolds for tissue engineering. *Rom. J. Morphol. Embryol.* **2018**, *59*, 71–76. [PubMed]
57. Cooper, C.; Rannou, F.; Richette, P.; Bruyere, O.; Al-Daghri, N.; Altman, R.D.; Brandi, M.L.; Collaud Basset, S.; Herrero-Beaumont, G.; Migliore, A.; et al. Use of intraarticular hyaluronic acid in the management of knee osteoarthritis in clinical practice. *Arthritis Care Res.* **2017**, *69*, 1287–1296. [CrossRef] [PubMed]
58. Stirma, G.A.; Chaves, D.H.; Tortato, S.; Belangero, P.S.; Lara, P.H.S.; Ejnisman, B. Prospective evaluation of periarticular hyaluronic acid infiltration for the treatment of lateral epicondylitis. *Acta Ortop. Bras.* **2020**, *28*, 107–110. [CrossRef] [PubMed]
59. Isa, I.L.; Srivastava, A.; Tiernan, D.; Owens, P.; Rooney, P.; Dockery, P.; Pandit, A. Hyaluronic acid based hydrogels attenuate inflammatory receptors and neurotrophins in interleukin-1beta induced inflammation model of nucleus pulposus cells. *Biomacromolecules* **2015**, *16*, 1714–1725. [CrossRef] [PubMed]
60. Kazezian, Z.; Li, Z.; Alini, M.; Grad, S.; Pandit, A. Injectable hyaluronic acid down-regulates interferon signaling molecules, IGFBP3 and IFIT3 in the bovine intervertebral disc. *Acta Biomater.* **2017**, *52*, 118–129. [CrossRef] [PubMed]
61. Murai, T. Lipid raft-mediated regulation of hyaluronan-CD44 interactions in inflammation and cancer. *Immunol. Front. Immunol.* **2015**, *6*, 420. [CrossRef] [PubMed]
62. Toole, B.P. Hyaluronan in morphogenesis. *J. Intern. Med.* **1997**, *242*, 35–40. [CrossRef] [PubMed]
63. Preston, M.; Sherman, L.S. Neural stem cell niches: Roles for the hyaluronan-based extracellular matrix. *Front. Biosci.* **2011**, *3*, 1165–1179. [CrossRef] [PubMed]
64. Lagendijk, A.K.; Szabo, A.; Merks, R.M.; Bakkers, J. Hyaluronan: A critical regulator of endothelial-to-mesenchymal transition during cardiac valve formation. *Trends Cardiovasc. Med.* **2013**, *23*, 135–142. [CrossRef]
65. Bastow, E.R.; Byers, S.; Golub, S.B.; Clarkin, C.E.; Pitsillides, A.A.; Fosang, A.J. Hyaluronan synthesis and degradation in cartilage and bone. *Cell Mol. Life Sci.* **2008**, *65*, 395–413. [CrossRef]
66. Bonnet, F.; Dunham, D.G.; Hardingham, T.E. Structure and interactions of cartilage proteoglycan binding region and link protein. *Biochem. J.* **1985**, *228*, 77–85. [CrossRef]
67. Fakhari, A.; Berkland, C. Applications and emerging trends of hyaluronic acid in tissue engineering, as a dermal filler and in osteoarthritis treatment. *Acta Biomater.* **2013**, *9*, 7081–7092. [CrossRef]
68. Hascall, V.; Esko, J.D. Hyaluronan. In *Essentials of Glycobiology*, 2nd ed.; Varki, A., Cummings, R.D., Esko, J.D., Freeze, H.H., Stanley, P., Bertozzi, C.R., Hart, G.W., Etzler, M.E., Eds.; Cold Spring Harbor (NY): New York, NY, USA, 2009.
69. Stern, R.; Asari, A.A.; Sugahara, K.N. Hyaluronan fragments: An information-rich system. *Eur. J. Cell Biol.* **2006**, *85*, 699–715. [CrossRef]
70. Moshayedi, P.; Carmichael, S.T. Hyaluronan, neural stem cells and tissue reconstruction after acute ischemic stroke. *Biomatter* **2013**, *3*, e23863. [CrossRef]
71. Stern, R.; Kogan, G.; Jedrzejas, M.J.; Soltes, L. The many ways to cleave hyaluronan. *Biotech. Adv.* **2007**, *25*, 537–557. [CrossRef]
72. Safrankova, B.; Gajdova, S.; Kubala, L. The potency of hyaluronan of different molecular weights in the stimulation of blood phagocytes. *Mediat. Inflamm.* **2010**, *2010*, 380948. [CrossRef] [PubMed]

73. Mukhopadhyay, D.; Asari, A.; Rugg, M.S.; Day, A.J.; Fulop, C. Specificity of the tumor necrosis factor-induced protein 6-mediated heavy chain transfer from inter-alpha-trypsin inhibitor to hyaluronan: Implications for the assembly of the cumulus extracellular matrix. *J. Biol. Chem.* **2004**, *279*, 11119–11128. [CrossRef] [PubMed]
74. Toole, B.P. Hyaluronan in morphogenesis. *Semin. Cell Dev. Biol.* **2001**, *12*, 79–87. [CrossRef] [PubMed]
75. Toole, B.P. Hyaluronan: From extracellular glue to pericellular cue. *Nat. Rev. Cancer* **2004**, *4*, 528–539. [CrossRef]
76. Collins, M.N.; Birkinshaw, C. Hyaluronic acid based scaffolds for tissue engineering—A review. *Carb Polym.* **2013**, *92*, 1262–1279. [CrossRef] [PubMed]
77. Strauss, E.J.; Hart, J.A.; Miller, M.D.; Altman, R.D.; Rosen, J.E. Hyaluronic acid viscosupplementation and osteoarthritis: Current uses and future directions. *Am. J. Sports Med.* **2009**, *37*, 1636–1644. [CrossRef] [PubMed]
78. Turley, E.A.; Noble, P.W.; Bourguignon, L.Y. Signaling properties of hyaluronan receptors. *J. Biol. Chem.* **2002**, *277*, 4589–4592. [CrossRef]
79. Fitzgerald, K.A.; Bowie, A.G.; Skeffington, B.S.; O'Neill, L.A. Ras, protein kinase C zeta, and I kappa B kinases 1 and 2 are downstream effectors of CD44 during the activation of NF-kappa B by hyaluronic acid fragments in T-24 carcinoma cells. *J. Immunol.* **2000**, *164*, 2053–2063. [CrossRef]
80. Hall, C.L.; Turley, E.A. Hyaluronan: RHAMM mediated cell locomotion and signaling in tumorigenesis. *J. Neurooncol.* **1995**, *26*, 221–229. [CrossRef]
81. Bollyky, P.L.; Bogdani, M.; Bollyky, J.B.; Hull, R.L.; Wight, T.N. The role of hyaluronan and the extracellular matrix in islet inflammation and immune regulation. *Curr. Diabetes Rep.* **2012**, *12*, 471–480. [CrossRef]
82. Wang, Q.G.; El Haj, A.J.; Kuiper, N.J. Glycosaminoglycans in the pericellular matrix of chondrons and chondrocytes. *J. Anat.* **2008**, *213*, 266–273. [CrossRef] [PubMed]
83. Pauloin, T.; Dutot, M.; Warnet, J.M.; Rat, P. In vitro modulation of preservative toxicity: High molecular weight hyaluronan decreases apoptosis and oxidative stress induced by benzalkonium chloride. *Eur. J. Pharm. Sci.* **2008**, *34*, 263–273. [CrossRef] [PubMed]
84. Taylor, K.R.; Trowbridge, J.M.; Rudisill, J.A.; Termeer, C.C.; Simon, J.C.; Gallo, R.L. Hyaluronan fragments stimulate endothelial recognition of injury through TLR4. *J. Biol. Chem.* **2004**, *279*, 17079–17084. [CrossRef] [PubMed]
85. Quero, L.; Klawitter, M.; Schmaus, A.; Rothley, M.; Sleeman, J.; Tiaden, A.N.; Klasen, J.; Boos, N.; Hottiger, M.O.; Wuertz, K.; et al. Hyaluronic acid fragments enhance the inflammatory and catabolic response in human intervertebral disc cells through modulation of toll-like receptor 2 signalling pathways. *Arthritis Res. Ther.* **2013**, *15*, R94. [CrossRef]
86. Urban, J.P.; Roberts, S. Degeneration of the intervertebral disc. *Arthritis Res. Ther.* **2003**, *5*, 120–130. [CrossRef] [PubMed]
87. Gullbrand, S.E.; Malhotra, N.R.; Schaer, T.P.; Zawacki, Z.; Martin, J.T.; Bendigo, J.R.; Milby, A.H.; Dodge, G.R.; Vresilovic, E.J.; Elliott, D.M.; et al. A large animal model that recapitulates the spectrum of human intervertebral disc degeneration. *Osteoarthr. Cartil.* **2017**, *25*, 146–156. [CrossRef]
88. Shi, C.; Qiu, S.; Riester, S.M.; Das, V.; Zhu, B.; Wallace, A.A.; van Wijnen, A.J.; Mwale, F.; Iatridis, J.C.; Sakai, D.; et al. Animal models for studying the etiology and treatment of low back pain. *J. Orthop. Res.* **2018**, *36*, 1305–1312. [CrossRef]
89. Hammer, R.E.; Maika, S.D.; Richardson, J.A.; Tang, J.P.; Taurog, J.D. Spontaneous inflammatory disease in transgenic rats expressing HLA-B27 and human beta 2m: An animal model of HLA-B27-associated human disorders. *Cell* **1990**, *63*, 1099–1112. [CrossRef]
90. Issy, A.C.; Castania, V.; Castania, M.; Salmon, C.E.; Nogueira-Barbosa, M.H.; Bel, E.D.; Defino, H.L. Experimental model of intervertebral disc degeneration by needle puncture in wistar rats. *Braz. J. Med. Biol. Res.* **2013**, *46*, 235–244. [CrossRef]
91. Sobajima, S.; Kompel, J.F.; Kim, J.S.; Wallach, C.J.; Robertson, D.D.; Vogt, M.T.; Kang, J.D.; Gilbertson, L.G. A slowly progressive and reproducible animal model of intervertebral disc degeneration characterized by MRI, X-ray, and histology. *Spine* **2005**, *30*, 15–24. [CrossRef]
92. Yurube, T.; Takada, T.; Suzuki, T.; Kakutani, K.; Maeno, K.; Doita, M.; Kurosaka, M.; Nishida, K. Rat tail static compression model mimics extracellular matrix metabolic imbalances of matrix metalloproteinases, aggrecanases, and tissue inhibitors of metalloproteinases in intervertebral disc degeneration. *Arthritis Res. Ther.* **2012**, *14*, R51. [CrossRef] [PubMed]

93. Alini, M.; Eisenstein, S.M.; Ito, K.; Little, C.; Kettler, A.A.; Masuda, K.; Melrose, J.; Ralphs, J.; Stokes, I.; Wilke, H.J. Are animal models useful for studying human disc disorders/degeneration? *Eur. Spine J.* **2008**, *17*, 2–19. [CrossRef] [PubMed]
94. Houseman, C.; Chen, D.; Scro, M.; Grande, D.A.; Levine, M. Chahine NO Treatment of intervertebral disc with chondroitinase-abc results in reversible degeneration in rat tail model. *ASME Summer Bioeng. Conf.* **2010**, *44038*, 853–854.
95. Deml, M.C.; Benneker, L.M.; Schmid, T.; Sakai, D.; Albers, C.E.; Hoppe, S.; Zeiter, S. Ventral surgical approach for an intervertebral disc degeneration and regeneration model in sheep cervical spine: Anatomic technical description, strengths and limitations. *Vet. Comp. Orthop. Traumatol.* **2019**, *32*, 389–393. [CrossRef] [PubMed]
96. Liang, Q.Q.; Zhou, Q.; Zhang, M.; Hou, W.; Cui, X.J.; Li, C.G.; Li, T.F.; Shi, Q.; Wang, Y.J. Prolonged upright posture induces degenerative changes in intervertebral discs in rat lumbar spine. *Spine* **2008**, *33*, 2052–2058. [CrossRef] [PubMed]
97. Wuertz, K.; Godburn, K.; MacLean, J.J.; Barbir, A.; Donnelly, J.S.; Roughley, P.J.; Alini, M.; Iatridis, J.C. In vivo remodeling of intervertebral discs in response to short- and long-term dynamic compression. *J. Orthop. Res.* **2009**, *27*, 1235–1242. [CrossRef]
98. Hamamoto, H.; Miyamoto, H.; Doita, M.; Takada, T.; Nishida, K.; Kurosaka, M. Capability of nondegenerated and degenerated discs in producing inflammatory agents with or without macrophage interaction. *Spine* **2012**, *37*, 161–167. [CrossRef]
99. Grivas, T.B.; Vasiliadis, E.S.; Kaspiris, A.; Khaldi, L.; Kletsas, D. Expression of matrix metalloproteinase-1 (MMP-1) in wistar rat's intervertebral disc after experimentally induced scoliotic deformity. *Scoliosis* **2011**, *6*, 9. [CrossRef]
100. Yuan, W.; Che, W.; Jiang, Y.Q.; Yuan, F.L.; Wang, H.R.; Zheng, G.L.; Li, X.L.; Dong, J. Establishment of intervertebral disc degeneration model induced by ischemic sub-endplate in rat tail. *Spine J.* **2015**, *15*, 1050–1059. [CrossRef]
101. Cuellar, J.M.; Borges, P.M.; Cuellar, V.G.; Yoo, A.; Scuderi, G.J.; Yeomans, D.C. Cytokine expression in the epidural space: A model of noncompressive disc herniation-induced inflammation. *Spine* **2013**, *38*, 17–23. [CrossRef]
102. Hirata, H.; Yurube, T.; Kakutani, K.; Maeno, K.; Takada, T.; Yamamoto, J.; Kurakawa, T.; Akisue, T.; Kuroda, R.; Kurosaka, M.; et al. A rat tail temporary static compression model reproduces different stages of intervertebral disc degeneration with decreased notochordal cell phenotype. *J. Orthop. Res.* **2014**, *32*, 455–463. [CrossRef] [PubMed]
103. Menard, A.L.; Grimard, G.; Massol, E.; Londono, I.; Moldovan, F.; Villemure, I. Static and dynamic compression application and removal on the intervertebral discs of growing rats. *J. Orthop. Res.* **2016**, *34*, 290–298. [CrossRef] [PubMed]
104. Han, C.; Ma, X.L.; Wang, T.; Ma, J.X.; Tian, P.; Zang, J.C.; Kong, J.B.; Li, X.D. Low magnitude of tensile stress represses the inflammatory response at intervertebral disc in rats. *J. Orthop. Surg. Res.* **2015**, *10*, 26. [CrossRef] [PubMed]
105. Hughes, P.C.; Tanner, J.M. The assessment of skeletal maturity in the growing rat. *J. Anat.* **1970**, *106 Pt 2*, 371–402.
106. Wang, F.; Gao, Z.X.; Cai, F.; Sinkemani, A.; Xie, Z.Y.; Shi, R.; Wei, J.N.; Wu, X.T. Formation, function, and exhaustion of notochordal cytoplasmic vacuoles within intervertebral disc: Current understanding and speculation. *Oncotarget* **2017**, *8*, 57800–57812. [CrossRef]
107. Iatridis, J.C.; Mente, P.L.; Stokes, I.A.; Aronsson, D.D.; Alini, M. Compression-induced changes in intervertebral disc properties in a rat tail model. *Spine* **1999**, *24*, 996–1002. [CrossRef]
108. Yurube, T.; Hirata, H.; Kakutani, K.; Maeno, K.; Takada, T.; Zhang, Z.; Takayama, K.; Matsushita, T.; Kuroda, R.; Kurosaka, M.; et al. Notochordal cell disappearance and modes of apoptotic cell death in a rat tail static compression-induced disc degeneration model. *Arthritis Res. Ther.* **2014**, *16*, R31. [CrossRef]
109. Guehring, T.; Omlor, G.W.; Lorenz, H.; Bertram, H.; Steck, E.; Richter, W.; Carstens, C.; Kroeber, M. Stimulation of gene expression and loss of anular architecture caused by experimental disc degeneration—An in vivo animal study. *Spine* **2005**, *30*, 2510–2515. [CrossRef]
110. Omlor, G.W.; Lorenz, H.; Engelleiter, K.; Richter, W.; Carstens, C.; Kroeber, M.W.; Guehring, T. Changes in gene expression and protein distribution at different stages of mechanically induced disc degeneration—An in vivo study on the New Zealand white rabbit. *J. Orthop. Res.* **2006**, *24*, 385–392. [CrossRef]

111. Chen, S.; Liu, S.; Ma, K.; Zhao, L.; Lin, H.; Shao, Z. TGF-beta signaling in intervertebral disc health and disease. *Osteoarthr. Cartil.* **2019**, *27*, 1109–1117. [CrossRef]
112. Yurube, T.; Ito, M.; Kakiuchi, Y.; Kuroda, R.; Kakutani, K. Autophagy and mTOR signaling during intervertebral disc aging and degeneration. *JOR Spine* **2020**, *3*, e1082. [CrossRef] [PubMed]
113. Kondo, N.; Yuasa, T.; Shimono, K.; Tung, W.; Okabe, T.; Yasuhara, R.; Pacifici, M.; Zhang, Y.; Iwamoto, M.; Enomoto-Iwamoto, M. Intervertebral disc development is regulated by Wnt/beta-catenin signaling. *Spine* **2011**, *36*, E513–E518. [CrossRef]
114. Smolders, L.A.; Meij, B.P.; Riemers, F.M.; Licht, R.; Wubbolts, R.; Heuvel, D.; Grinwis, G.C.; Vernooij, H.C.; Hazewinkel, H.A.; Penning, L.C.; et al. Canonical Wnt signaling in the notochordal cell is upregulated in early intervertebral disk degeneration. *J. Orthop. Res.* **2012**, *30*, 950–957. [CrossRef] [PubMed]
115. Erwin, W.M.; DeSouza, L.; Funabashi, M.; Kawchuk, G.; Karim, M.Z.; Kim, S.; Mdler, S.; Matta, A.; Wang, X.; Mehrkens, K.A. The biological basis of degenerative disc disease: Proteomic and biomechanical analysis of the canine intervertebral disc. *Arthritis Res. Ther.* **2015**, *17*, 240. [CrossRef]
116. Ukita, K.; Hirahara, S.; Oshima, N.; Imuta, Y.; Yoshimoto, A.; Jang, C.W.; Oginuma, M.; Saga, Y.; Behringer, R.R.; Kondoh, H.; et al. Wnt signaling maintains the notochord fate for progenitor cells and supports the posterior extension of the notochord. *Mech. Dev.* **2009**, *126*, 791–803. [CrossRef] [PubMed]
117. Wang, M.; Tang, D.; Shu, B.; Wang, B.; Jin, H.; Hao, S.; Dresser, K.A.; Shen, J.; Im, H.J.; Sampson, E.R.; et al. Conditional activation of beta-catenin signaling in mice leads to severe defects in intervertebral disc tissue. *Arthritis Rheum.* **2012**, *64*, 2611–2623. [CrossRef] [PubMed]
118. Holguin, N.; Silva, M.J. In-vivo nucleus pulposus-specific regulation of adult murine intervertebral disc degeneration via wnt/beta-catenin signaling. *Sci. Rep.* **2018**, *8*, 11191. [CrossRef]
119. Fujita, N.; Gogate, S.S.; Chiba, K.; Toyama, Y.; Shapiro, I.M.; Risbud, M.V. Prolyl hydroxylase 3 (PHD3) modulates catabolic effects of tumor necrosis factor-alpha (TNF-alpha) on cells of the nucleus pulposus through co-activation of nuclear factor kappaB (NF-kappaB)/p65 signaling. *J. Biol. Chem.* **2012**, *287*, 39942–39953. [CrossRef]
120. Wako, M.; Ohba, T.; Ando, T.; Arai, Y.; Koyama, K.; Hamada, Y.; Nakao, A.; Haro, H. Mechanism of signal transduction in tumor necrosis factor-like weak inducer of apoptosis-induced matrix degradation by MMP-3 upregulation in disc tissues. *Spine* **2008**, *33*, 2489–2494. [CrossRef]
121. Hoyland, J.A.; Le Maitre, C.; Freemont, A.J. Investigation of the role of IL-1 and TNF in matrix degradation in the intervertebral disc. *Rheumatology* **2008**, *47*, 809–814. [CrossRef]
122. Wang, J.; Markova, D.; Anderson, D.G.; Zheng, Z.; Shapiro, I.M.; Risbud, M.V. TNF-alpha and IL-1beta promote a disintegrin-like and metalloprotease with thrombospondin type I motif-5-mediated aggrecan degradation through syndecan-4 in intervertebral disc. *J. Biol. Chem.* **2011**, *286*, 39738–39749. [CrossRef] [PubMed]
123. Ohba, T.; Haro, H.; Ando, T.; Wako, M.; Suenaga, F.; Aso, Y.; Koyama, K.; Hamada, Y.; Nakao, A. TNF-alpha-induced NF-kappaB signaling reverses age-related declines in VEGF induction and angiogenic activity in intervertebral disc tissues. *J. Orthop. Res.* **2009**, *27*, 229–235. [CrossRef] [PubMed]
124. Liang, H.; Yang, X.; Liu, C.; Sun, Z.; Wang, X. Effect of NF-kB signaling pathway on the expression of MIF, TNF-alpha, IL-6 in the regulation of intervertebral disc degeneration. *J. Musculoskelet. Neuronal Interact.* **2018**, *18*, 551–556. [PubMed]
125. Gao, G.; Chang, F.; Zhang, T.; Huang, X.; Yu, C.; Hu, Z.; Ji, M.; Duan, Y. Naringin protects against interleukin 1beta (IL-1beta)-Induced human nucleus pulposus cells degeneration via downregulation nuclear factor kappa B (NF-kappaB) pathway and p53 expression. *Med. Sci. Monit.* **2019**, *25*, 9963–9972. [CrossRef]
126. Gruber, H.E.; Hoelscher, G.L.; Bethea, S.; Hanley, E.N., Jr. Interleukin 1-beta up-regulates brain-derived neurotrophic factor, neurotrophin 3 and neuropilin 2 gene expression and NGF production in annulus cells. *Biotech. Histochem.* **2012**, *87*, 506–511. [CrossRef]
127. Lee, J.M.; Song, J.Y.; Baek, M.; Jung, H.Y.; Kang, H.; Han, I.B.; Kwon, Y.D.; Shin, D.E. Interleukin-1beta induces angiogenesis and innervation in human intervertebral disc degeneration. *J. Orthop. Res.* **2011**, *29*, 265–269. [CrossRef]
128. Richardson, S.M.; Doyle, P.; Minogue, B.M.; Gnanalingham, K.; Hoyland, J.A. Increased expression of matrix metalloproteinase-10, nerve growth factor and substance P in the painful degenerate intervertebral disc. *Arthritis Res. Ther.* **2009**, *11*, R126. [CrossRef]

129. Yamauchi, K.; Inoue, G.; Koshi, T.; Yamashita, M.; Ito, T.; Suzuki, M.; Eguchi, Y.; Orita, S.; Takaso, M.; Nakagawa, K.; et al. Nerve growth factor of cultured medium extracted from human degenerative nucleus pulposus promotes sensory nerve growth and induces substance p in vitro. *Spine* **2009**, *34*, 2263–2269. [CrossRef]
130. Freemont, A.J.; Watkins, A.; Le Maitre, C.; Baird, P.; Jeziorska, M.; Knight, M.T.; Ross, E.R.; O'Brien, J.P.; Hoyland, J.A. Nerve growth factor expression and innervation of the painful intervertebral disc. *J. Pathol.* **2002**, *197*, 286–292. [CrossRef]
131. Abe, Y.; Akeda, K.; An, H.S.; Aoki, Y.; Pichika, R.; Muehleman, C.; Kimura, T.; Masuda, K. Pro-inflammatoryry cytokines stimulate the expression of nerve growth factor by human intervertebral disc cells. *Spine* **2007**, *32*, 635–642. [CrossRef]
132. Pratsinis, H.; Constantinou, V.; Pavlakis, K.; Sapkas, G.; Kletsas, D. Exogenous and autocrine growth factors stimulate human intervertebral disc cell proliferation via the ERK and Akt pathways. *J. Orthop. Res.* **2012**, *30*, 958–964. [CrossRef] [PubMed]
133. Mavrogonatou, E.; Kletsas, D. Effect of varying osmotic conditions on the response of bovine nucleus pulposus cells to growth factors and the activation of the ERK and Akt pathways. *J. Orthop. Res.* **2010**, *28*, 1276–1282. [CrossRef] [PubMed]
134. Studer, R.K.; Gilbertson, L.G.; Georgescu, H.; Sowa, G.; Vo, N.; Kang, J.D. p38 MAPK inhibition modulates rabbit nucleus pulposus cell response to IL-1. *J. Orthop. Res.* **2008**, *26*, 991–998. [CrossRef] [PubMed]
135. Tsai, T.T.; Guttapalli, A.; Agrawal, A.; Albert, T.J.; Shapiro, I.M.; Risbud, M.V. MEK/ERK signaling controls osmoregulation of nucleus pulposus cells of the intervertebral disc by transactivation of TonEBP/OREBP. *J. Bone Miner. Res.* **2007**, *22*, 965–974. [CrossRef] [PubMed]
136. Seguin, C.A.; Bojarski, M.; Pilliar, R.M.; Roughley, P.J.; Kandel, R.A. Differential regulation of matrix degrading enzymes in a TNFalpha-induced model of nucleus pulposus tissue degeneration. *Matrix Biol.* **2006**, *25*, 409–418. [CrossRef]
137. Risbud, M.V.; Fertala, J.; Vresilovic, E.J.; Albert, T.J.; Shapiro, I.M. Nucleus pulposus cells up-regulate PI3K/Akt and MEK/ERK signaling pathways under hypoxic conditions and resist apoptosis induced by serum withdrawal. *Spine* **2005**, *30*, 882–889. [CrossRef]
138. Risbud, M.V.; Guttapalli, A.; Albert, T.J.; Shapiro, I.M. Hypoxia activates MAPK activity in rat nucleus pulposus cells: Regulation of integrin expression and cell survival. *Spine* **2005**, *30*, 2503–2509. [CrossRef]
139. Hiyama, A.; Skubutyte, R.; Markova, D.; Anderson, D.G.; Yadla, S.; Sakai, D.; Mochida, J.; Albert, T.J.; Shapiro, I.M.; Risbud, M.V. Hypoxia activates the notch signaling pathway in cells of the intervertebral disc: Implications in degenerative disc disease. *Arthritis Rheum.* **2011**, *63*, 1355–1364. [CrossRef]
140. Tsai, T.T.; Guttapalli, A.; Oguz, E.; Chen, L.H.; Vaccaro, A.R.; Albert, T.J.; Shapiro, I.M.; Risbud, M.V. Fibroblast growth factor-2 maintains the differentiation potential of nucleus pulposus cells in vitro: Implications for cell-based transplantation therapy. *Spine* **2007**, *32*, 495–502. [CrossRef]
141. Li, X.; An, H.S.; Ellman, M.; Phillips, F.; Thonar, E.J.; Park, D.K.; Udayakumar, R.K.; Im, H.J. Action of fibroblast growth factor-2 on the intervertebral disc. *Arthritis Res. Ther.* **2008**, *10*, R48. [CrossRef]
142. Alini, M.; Li, W.; Markovic, P.; Aebi, M.; Spiro, R.C.; Roughley, P.J. The potential and limitations of a cell-seeded collagen/hyaluronan scaffold to engineer an intervertebral disc-like matrix. *Spine* **2003**, *28*, 446–454; discussion 453. [CrossRef]
143. Nakashima, S.; Matsuyama, Y.; Takahashi, K.; Satoh, T.; Koie, H.; Kanayama, K.; Tsuji, T.; Maruyama, K.; Imagama, S.; Sakai, Y.; et al. Regeneration of intervertebral disc by the intradiscal application of cross-linked hyaluronate hydrogel and cross-linked chondroitin sulfate hydrogel in a rabbit model of intervertebral disc injury. *Biomed. Mater. Eng.* **2009**, *19*, 421–429. [CrossRef]
144. Pfeiffer, M.; Boudriot, U.; Pfeiffer, D.; Ishaque, N.; Goetz, W.; Wilke, A. Intradiscal application of hyaluronic acid in the non-human primate lumbar spine: Radiological results. *Eur. Spine J.* **2003**, *12*, 76–83. [CrossRef] [PubMed]
145. Kazezian, Z.; Sakai, D.; Pandit, A. Hyaluronic acid microgels modulate inflammation and key matrix molecules toward a regenerative signature in the injured annulus fibrosus. *Adv. Biosyst.* **2017**, *1*, 1–14. [CrossRef] [PubMed]
146. Omlor, G.W.; Nerlich, A.G.; Lorenz, H.; Bruckner, T.; Richter, W.; Pfeiffer, M.; Guhring, T. Injection of a polymerised hyaluronic acid/collagen hydrogel matrix in an in vivo porcine disc degeneration model. *Eur. Spine J.* **2012**, *21*, 1700–1708. [CrossRef] [PubMed]

147. Henriksson, H.B.; Hagman, M.; Horn, M.; Lindahl, A.; Brisby, H. Investigation of different cell types and gel carriers for cell-based intervertebral disc therapy, in vitro and in vivo studies. *J. Tissue Eng. Regen. Med.* **2012**, *6*, 738–747. [CrossRef]
148. Mohd Isa, I.L.; Abbah, S.A.; Kilcoyne, M.; Sakai, D.; Dockery, P.; Finn, D.P.; Pandit, A. Implantation of hyaluronic acid hydrogel prevents the pain phenotype in a rat model of intervertebral disc injury. *Sci. Adv.* **2018**, *4*, eaaq0597. [CrossRef]
149. Pan, Z.; Sun, H.; Xie, B.; Xia, D.; Zhang, X.; Yu, D.; Li, J.; Xu, Y.; Wang, Z.; Wu, Y.; et al. Therapeutic effects of gefitinib-encapsulated thermosensitive injectable hydrogel in intervertebral disc degeneration. *Biomaterials* **2018**, *160*, 56–68. [CrossRef]
150. Hackel, S.; Zolfaghar, M.; Du, J.; Hoppe, S.; Benneker, L.M.; Garstka, N.; Peroglio, M.; Alini, M.; Grad, S.; Yayon, A.; et al. Fibrin-hyaluronic acid hydrogel (RegenoGel) with fibroblast growth factor-18 for in vitro 3D culture of human and bovine nucleus pulposus Cells. *Int. J. Mol. Sci.* **2019**, *20*, 5036. [CrossRef]
151. Watanabe, A.; Mainil-Varlet, P.; Decambron, A.; Aschinger, C.; Schiavinato, A. Efficacy of HYADD(R)4-G single intra-discal injections in a rabbit model of intervertebral disc degeneration. *Biomed. Mater. Eng.* **2019**, *30*, 403–417. [CrossRef]
152. Vadala, G.; Russo, F.; Musumeci, M.; D'Este, M.; Cattani, C.; Catanzaro, G.; Tirindelli, M.C.; Lazzari, L.; Alini, M.; Giordano, R.; et al. Clinically relevant hydrogel-based on hyaluronic acid and platelet rich plasma as a carrier for mesenchymal stem cells: Rheological and biological characterization. *J. Orthop. Res.* **2017**, *35*, 2109–2116. [CrossRef] [PubMed]
153. Fuller, E.S.; Shu, C.; Smith, M.M.; Little, C.B.; Melrose, J. Hyaluronan oligosaccharides stimulate matrix metalloproteinase and anabolic gene expression in vitro by intervertebral disc cells and annular repair in vivo. *J. Tissue Eng. Regen. Med.* **2018**, *12*, e216–e226. [CrossRef] [PubMed]
154. Peeters, M.; Detiger, S.E.; Karfeld-Sulzer, L.S.; Smit, T.H.; Yayon, A.; Weber, F.E.; Helder, M.N. BMP-2 and BMP-2/7 heterodimers conjugated to a fibrin/hyaluronic acid hydrogel in a large animal model of mild intervertebral disc degeneration. *Biores* **2015**, *4*, 398–406. [CrossRef] [PubMed]
155. Chen, P.; Ning, L.; Qiu, P.; Mo, J.; Mei, S.; Xia, C.; Zhang, J.; Lin, X.; Fan, S. Photo-crosslinked gelatin-hyaluronic acid methacrylate hydrogel-committed nucleus pulposus-like differentiation of adipose stromal cells for intervertebral disc repair. *J. Tissue Eng. Regen. Med.* **2019**, *13*, 682–693. [CrossRef] [PubMed]
156. Gansau, J.; Buckley, C.T. Incorporation of collagen and hyaluronic acid to enhance the bioactivity of fibrin-based hydrogels for nucleus pulposus regeneration. *J. Funct. Biomater.* **2018**, *9*, 43. [CrossRef]
157. Kumar, H.; Ha, D.H.; Lee, E.J.; Park, J.H.; Shim, J.H.; Ahn, T.K.; Kim, K.T.; Ropper, A.E.; Sohn, S.; Kim, C.H.; et al. Safety and tolerability of intradiscal implantation of combined autologous adipose-derived mesenchymal stem cells and hyaluronic acid in patients with chronic discogenic low back pain: 1-year follow-up of a phase I study. *Stem Cell Res. Ther.* **2017**, *8*, 262. [CrossRef]
158. Fuchs, S.; Erbe, T.; Fischer, H.L.; Tibesku, C.O. Intraarticular hyaluronic acid versus glucocorticoid injections for nonradicular pain in the lumbar spine. *J. Vasc. Interv. Radiol.* **2005**, *16*, 1493–1498. [CrossRef]
159. Termeer, C.; Benedix, F.; Sleeman, J.; Fieber, C.; Voith, U.; Ahrens, T.; Miyake, K.; Freudenberg, M.; Galanos, C.; Simon, J.C. Oligosaccharides of hyaluronan activate dendritic cells via toll-like receptor 4. *J. Exp. Med.* **2002**, *195*, 99–111. [CrossRef]
160. Nesti, L.J.; Li, W.J.; Shanti, R.M.; Jiang, Y.J.; Jackson, W.; Freedman, B.A.; Kuklo, T.R.; Giuliani, J.R.; Tuan, R.S. Intervertebral disc tissue engineering using a novel hyaluronic acid-nanofibrous scaffold (HANFS) amalgam. *Tissue Eng.* **2008**, *14*, 1527–1537. [CrossRef]
161. Safety and Preliminary Efficacy Study Of Mesenchymal Precursor Cells (MPCs) in Subjects with Lumbar Back Pain. Available online: https://clinicaltrials.gov/ct2/show/NCT01290367 (accessed on 10 August 2020).
162. Thones, S.; Rother, S.; Wippold, T.; Blaszkiewicz, J.; Balamurugan, K.; Moeller, S.; Ruiz-Gomez, G.; Schnabelrauch, M.; Scharnweber, D.; Saalbach, A.; et al. Hyaluronan/collagen hydrogels containing sulfated hyaluronan improve wound healing by sustained release of heparin-binding EGF-like growth factor. *Acta Biomater.* **2019**, *86*, 135–147. [CrossRef]
163. Joy, R.A.; Vikkath, N.; Ariyannur, P.S. Metabolism and mechanisms of action of hyaluronan in human biology. *Drug Metab. Pers. Ther.* **2018**, *33*, 15–32. [CrossRef] [PubMed]
164. Melrose, J.; Shu, C.; Young, C.; Ho, R.; Smith, M.M.; Young, A.A.; Smith, S.S.; Gooden, B.; Dart, A.; Podadera, J.; et al. Mechanical destabilization induced by controlled annular incision of the intervertebral disc dysregulates metalloproteinase expression and induces disc degeneration. *Spine* **2012**, *37*, 18–25. [CrossRef] [PubMed]

165. Colloca, C.J.; Gunzburg, R.; Freeman, B.J.; Szpalski, M.; Afifi, M.; Moore, R.J. Biomechancial quantification of pathologic manipulable spinal lesions: An in vivo ovine model of spondylolysis and intervertebral disc degeneration. *J. Manipul. Physiol. Ther.* **2012**, *35*, 354–366. [CrossRef] [PubMed]
166. Imai, Y.; Okuma, M.; An, H.S.; Nakagawa, K.; Yamada, M.; Muehleman, C.; Thonar, E.; Masuda, K. Restoration of disc height loss by recombinant human osteogenic protein-1 injection into intervertebral discs undergoing degeneration induced by an intradiscal injection of chondroitinase ABC. *Spine* **2007**, *32*, 1197–1205. [CrossRef]
167. Petrey, A.C.; de la Motte, C.A. Hyaluronan, a crucial regulator of inflammation. *Front. Immunol.* **2014**, *5*, 101. [CrossRef]
168. Misra, S.; Hascall, V.C.; Markwald, R.R.; Ghatak, S. Interactions between hyaluronan and its receptors (CD44, RHAMM) regulate the activities of inflammation and cancer. *Front. Immunol.* **2015**, *6*, 201. [CrossRef]

Article

In Vitro Model for Lumbar Disc Herniation to Investigate Regenerative Tissue Repair Approaches

Laura Zengerle [1], Elisabeth Debout [1], Bruno Kluger [2], Lena Zöllner [1] and Hans-Joachim Wilke [1,*]

[1] Institute of Orthopaedic Research and Biomechanics, Ulm University, 89081 Ulm, Germany; laura.zengerle@uni-ulm.de (L.Z.); elisabeth.debout@mail.de (E.D.); lena.zoellner@uni-ulm.de (L.Z.)
[2] Institute for Laser Technologies in Medicine and Metrology, Ulm University, 89081 Ulm, Germany; brunokluger@gmx.de
* Correspondence: hans-joachim.wilke@uni-ulm.de

Featured Application: Development of a lumbar disc herniation model with which regenerative tissue repair approaches can be investigated under physiological loading conditions and worst-case scenarios out of patients' daily-life.

Abstract: Lumbar disc herniation (LDH) is the most common reason for low back pain in the working society. New regenerative approaches and novel implants are directed towards the restoration of the disc or its biomechanical properties. Aiming to investigate these new therapies under physiological conditions, in this study, a model for LDH was established by developing a new physiological in vitro test method. In 14 human lumbar motion segments, different daily-life and worst-case activities were simulated successfully by applying a physiological range of motion and axial loading in order to create physiological intradiscal pressure. An LDH could be provoked in 11 of the 14 specimens through vertical and round annular defects of different sizes. Interestingly, the defect and the LDH hardly influenced the biomechanical properties of the disc. For the investigation of regenerative approaches in further experiments, the recommendation based on the results of this study is to create an LDH in non-degenerated motion segments by the application of the new physiological in vitro test method after setting the round annular defects to a size of 4 mm in diameter.

Keywords: test method; dynamic testing; in vitro; (lumbar) disc herniation; physiological activities; regenerative approaches; tissue engineering

Citation: Zengerle, L.; Debout, E.; Kluger, B.; Zöllner, L.; Wilke, H.-J. In Vitro Model for Lumbar Disc Herniation to Investigate Regenerative Tissue Repair Approaches. *Appl. Sci.* **2021**, *11*, 2847. https://doi.org/10.3390/app11062847

Academic Editor: Benjamin Gantenbein

Received: 26 February 2021
Accepted: 21 March 2021
Published: 22 March 2021

1. Introduction

Lumbar disc herniation is the most common reason for low back pain in the working society [1]. The incidence ranges from 1 to 5 of 1000 persons per year but shows a peak in patients with an age ranging from 30 to 50 years [2]. The lower lumbar region is mostly affected with a proportion of over 95% of all lumbar disc herniations, whereas 50% occur in L4-5 and 46% in L5-S1 [3,4]. An intervertebral disc prolapse with extrusion of nucleus pulposus can lead to a compression of nerve roots which can cause pain, numbness and even paralysis.

For a long time, therapies targeted the reduction of these symptoms by surgical nerve decompression, often in combination with implants. New regenerative approaches and novel implants are directed towards the restoration of the disc itself with the goal to improve the biomechanical properties [5–7] and the biological functions of an intact disc [8–11]. This motivates the development of new materials as well as the use of innovative technologies such as 3D bioprinting which offers the opportunity to create highly complex and multi-dimensional biomaterials [12–14]. The investigation of such new treatment options also requires biomechanical experiments where a lumbar disc herniation can be provoked under realistic conditions in order to challenge the new treatments under physiological worst-case scenarios.

Therefore, it is of considerable interest to understand the mechanisms of when, where, and how lumbar disc herniations develop. It is known that most lumbar disc herniations result from endplate junction failures (65%) rather than an annulus fibrosus rupture (35%) [15,16]. However, which activities lead to the development of such lumbar disc herniations, are yet unknown. Activities that were associated with a higher risk for a gradual development of lumbar disc herniations include flexional movements or lifting weights [17–20]. Further, it has been assumed that complex and extreme loading conditions might facilitate disc herniations.

Various in vitro studies investigated the influence of these different loading conditions or motions on structural failures of the lumbar motion segments [2,15,21–30]. The results of those studies confirmed that the combination of bending forward in combination with lateral bending and axial rotation while lifting heavy objects is considered to be harmful [31–33]. Nevertheless, in most of these studies, an exaggerated range of motion (ROM) or loading values had to be applied in order to simulate the long-term effects expected from physiological movements [15,30]. So far, no herniation model could be created that allows the investigation of new regenerative therapies under physiological loading conditions and worst-case scenarios. Furthermore, the precise biomechanical mechanism of the development of lumbar disc herniations is still not completely understood. It is still uncertain during which physiologic activity a disc herniation most likely occurs.

Therefore, the aim of this study was to establish a lumbar disc herniation model by using a new biomechanical in vitro test method which simulates patients' daily life activities. With this new test method, lumbar disc herniations should be produced through standardized annular defects and dynamic physiological loading. This lumbar disc herniation model should allow the investigation of novel implants and regenerative therapies under worst-case loading scenarios that really happen during patients' daily life activities.

2. Materials and Methods

2.1. Specimens

In this study, we used 14 single lumbo-sacral motion segments (L2-3, L3-4, L4-5, L5-S1) from eight different adult human donors with a median age of 35.5 years, ranging from 19 to 53 years (Table 1). Ethical approval was provided by the ethical committee board. Sagittal T1 and T2-weighted and transverse T2-weighted magnetic resonance imaging (MRI) was performed to evaluate disc quality and to exclude specimens with disc damage such as disc herniations. Only discs with a low degree of disc degeneration (according to Pfirrmann 1 or 2) were chosen and included in this study. CT scans were performed to exclude specimens with bony fractures, tumors or other problems. The segments were then dissected removing all soft tissue and keeping the disc and all functional ligaments intact. The cranial and caudal vertebrae were embedded into polymethylmethacrylate (PMMA, Technovit 3040, Haereus Kulzer, Wehrheim, Germany) and flanges were mounted to enable a proper fixation in the testing machines.

In all specimens, a right lateral laminotomy was performed to allow access to and observation of the posterior annulus fibrosus with an endoscopic camera. Prior to the calibration of the physiological testing protocol, the specimens were hydrated in physiological saline solution for 1 h to ensure physiologically high hydrostatic pressure in the nucleus pulposus. Before dynamically loading the specimens with the new physiological test protocol, the specimens were rehydrated for 30 min [34] and kept moist throughout the experiment [35].

Table 1. Motion segments from human donors that were used for this study for the different defect groups. Only segments with hardly no or low degeneration (degree of degeneration according to Pfirrmann [36] 1–2) and no relevant diagnoses such as present disc herniation that led to exclusion of the specimen were chosen.

Defect Group	Donor	Sex	Age in Years	Segment	Degeneration (acc. to Pfirrmann [36])	Relevant Diagnoses
vertical cut	2130	male	26	L2-L3	2	
				L4-L5	3–4	disc herniation [1]
	2133	female	24	L2-L3	1	
				L4-L5	1	
	2134	male	26	L3-L4	1	
				L5-S1	4	disc herniation [1]
	2138	male	33	L2-L3	1	
				L4-L5	2	
	median (range)		26 (24, 33)		1 (1, 2) [2]	
round hole	2036	male	40	L2-L3	1	
				L4-L5	1	
	2072	male	53	L3-L4	1–2	
				L5-S1	1–2	
	2129	n.a.	19	L2-L3	1	
				L4-L5	1	
	2132	male	31	L2-L3	1	
				L4-L5	1	
	median (range)		35.5 (19, 53)		1 (1, 1–2)	
overall	median (range)		35.5 (19, 53)		1 (1, 2)	

[1] The diagnosis disc herniation led to exclusion of those specimens. [2] Median (range) based on specimens that were included in the study.

2.2. New Physiological Loading Protocol

A new physiological loading protocol was developed to simulate typical daily in vivo activities associated with lumbar disc herniations. Therefore, activities were chosen that either consist of extreme flexion like tying shoes or with additional loading like lifting boxes. Furthermore, the influence of activities with complex motion patterns on the risk of lumbar disc herniations was investigated by simulating sweeping the floor or lifting boxes while turning. For each activity, physiological in vivo segmental motion [37] and intradiscal pressures (IDP, [38]) were replicated in a dynamic disc loading simulator [39]. This dynamic disc loading simulator is able to deform the specimens in six degrees of freedom and thereby allows the replication of complex motion in all anatomical planes, as well as additional axial loading or compression to simulate resultant muscle forces or body weight. In order to monitor the IDP that results from this complex deformation and loading of the specimen, an IDP sensor (Mammendorfer Institut für Physik und Medizin GmbH, Hattenhofen, Germany) was implanted into the nucleus pulposus from the right lateral side of the disc.

From a neutral position, the starting activity of the patient in a standing position was approached. Then, the dynamic activities tying shoes, sweeping the floor, lifting boxes (20 kg) and the corresponding complex activity lifting boxes while turning were performed subsequently (Supplementary Materials: Video S1). Each activity consisted either of pure axial loading (standing) or in combination with bending. For the simulation of tying shoes and lifting boxes, the specimens were bent in flexion; for the simulation of sweeping the floor or lifting boxes while turning in flexion, lateral bending and axial rotation were applied simultaneously (Table 2). The motion values that were applied for each activity resulted from an in vivo study performed by Pearcy in 1985 [37]. The amount of the necessary axial load for every individual specimen was determined from in vivo IDP measurements from the study by Wilke et al. [38]. During a quasistatic calibration cycle prior to testing, the equivalent motion of each activity was applied and axial compression was subsequently adjusted (Table 2) until the equivalent in vivo IDP value had been created within the disc. However, in vivo IDP was known for the activities of standing, tying shoes, and lifting boxes. For the simulation of the complex activities (sweeping the floor and lifting boxes while turning), rotations in all motion directions were applied. The axial load was maintained with respect to the corresponding simple activity.

Table 2. Daily-life activities (*standing, tying shoes, sweeping floor, lifting boxes, lifting boxes while turning*) simulated by application of the equivalent physiological range of motion (ROM), exemplarily for L4-L5 [25] in the main bending directions, and specimen-specific axial loading in kN in order to achieve the intradiscal pressure (IDP) in MPa that was measured in vivo [26]. The resultant IDP peaks that could be measured during dynamic loading show deviations from the target IDP that was used for quasistatic calibration.

Activities	Standing	Tying Shoes	Sweeping Floor	Lifting Boxes	Lifting Boxes while Turning
bending directions (with ROM for L4-L5 [25])	neutral (0°)	flexion 13°	flexion (13°) left lateral bending (2°) left axial rotation (2°)	flexion 13°	flexion (13°) left lateral bending (2°) left axial rotation (2°)
IDP in MPa [26]	0.5	1.1	n.a.	2.3	n.a.
Axial load in kN	0.45 (0.24–0.54)	0.97 (0.24–2.66)	1.45 (0.23–2.63)	2.13 (1.14–3.57)	1.9 (1.13–3.58)
IDP peaks in MPa during dyn. loading	0.5 (0.4–0.8)	1.7 (1.4–3.5)	1.9 (1.4–3.3)	3.0 (2.7–3.6)	2.8 (1.9–3.2)

The dynamic loading protocol applied was motion controlled for all ranges of motion and force controlled for the axial compression in order to create the IDP. Each activity was simulated for six cycles at 0.1 Hz. IDP was monitored constantly during dynamic loading. The specimens were loaded first in the intact state of the intervertebral disc and then after setting annular defects.

2.3. Defects

In the posterior part of the annulus fibrosus, vertical (scalpel cut 0.4 mm × 4.0 mm, 1.0 mm × 5.5 mm, 1.2 mm × 6.5 mm) or round defects (Ø 4 mm, Ø 6 mm, Ø 8 mm) were created in the specimens and enlarged consecutively. The posterior annulus fibrosus was observed during the simulation of the activities using an endoscopic camera in order to evaluate whether and at which cycle of which activity a herniation with clear nucleus extrusion occurred. The system consists of a high-performance light unit (D-Light-N, 20t33420 Karl Storz SE & Co. KG, Tuttlingen, Germany), a fiber optic light transmission incorporated telescope (PA-NOVIEW, Richard Wolf GmbH, Knittlingen, Germany) and an ultra-compact USB 3.0 camera (xiQ, XIMEA GmbH, Münster, Germany). After occurrence of a herniation, the defect was not further enlarged. In every case a herniation occurred through a vertical defect; scans in an ultra-high field MRI (11.7 T, BioSpec 117/16, Bruker Corp., Billerica, MA, USA) were taken using isotropic voxels with a resolution of 100 μm.

2.4. Flexibility Tests

For comparison of the biomechanical properties of the specimens with intact, injured, or herniated disc, standardized torsion tests were carried out. Hence, quasistatic flexibility tests were performed in all testing conditions: in the intact disc, after setting the defects and after a herniation was provoked during dynamical loading by use of the new physiological loading protocol. Therefore, a universal spine loading simulator [40] was used to apply pure moments of ±7.5 Nm in flexion-extension, lateral bending, and axial rotation [41]. Maximum ROM and neutral zone (NZ) were measured by a motion tracking system (Vicon Nexus 1.4.116, Vicon Motion Systems Ltd., Oxford, UK) with six cameras (Type MX13, Vicon Motion Systems Ltd., Oxford, UK). Maximum IDP that was reached in the extrema of the flexibility test, was recorded by the same IDP sensor that was already implanted into the nucleus pulposus and that was also used for the calibration of the dynamic loading protocol.

2.5. Statistical Analysis

Statistical analysis was performed using a Friedman-Test with Bonferroni Post-Hoc correction in SPSS Software (IBM SPSS Statistics Vol. 27; IBM Corp., Armonk, NY, USA). The significance level was set to $\alpha \leq 0.05$.

3. Results

In this study, a new test method could be developed which simulates physiological daily-life activities in a dynamic way. In vivo motion values could be successfully replicated. By combining it with an axial compression, in vivo IDP could be created during the calibration cycle. During the dynamic validation cycle, the in vivo IDP values for *standing* were reached, but exceeded for *tying shoes* and *lifting boxes* (Table 2).

3.1. Provocation of Disc Herniation

With this method, a prolapse with clear nucleus extrusion (Figure 1a) could be provoked in 11 of the 14 specimens (Supplementary Materials: Video S2). A previously set defect was necessary in all specimens (Table 3).

Table 3. Lumbar disc herniations (LDH) that could be provoked in the specimens with different defects during the simulation of physiological activities indicating when the herniation occurred.

Disc Condition	Defect Size	Number of Herniations	Physiological Activity (Cycle) When LDH Occurred
intact disc	-	0	data
vertical defect	0.4 mm × 4.0 mm	1	 (*n* = 1)
vertical defect	1.0 mm × 5.5 mm	1	 (*n* = 1)
vertical defect	1.2 mm × 6.5 mm	3	 (*n* = 1)
			 (*n* = 2)
One specimen with vertical defect did NOT herniate			
round defect	Ø 4 mm	6	before dynamic loading (*n* = 3) [1]  (*n* = 3)
round defect	Ø 6 mm	no further LDH †	
round defect	Ø 8 mm	no further LDH †	

[1] In three specimens, the LDH already occurred during the flexibility test (under pure moments with no preload) before dynamic loading.
† Two specimens with round defect did NOT herniate at all, even with 8 mm.

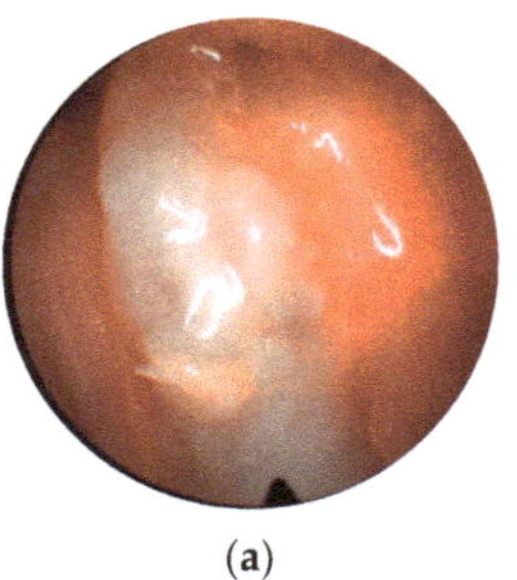
(a)

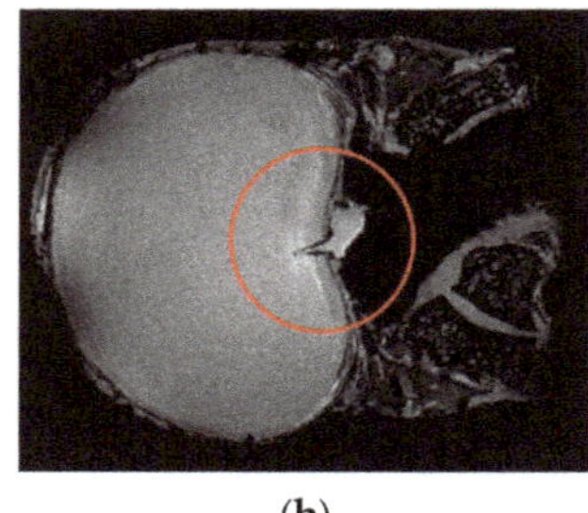
(b)

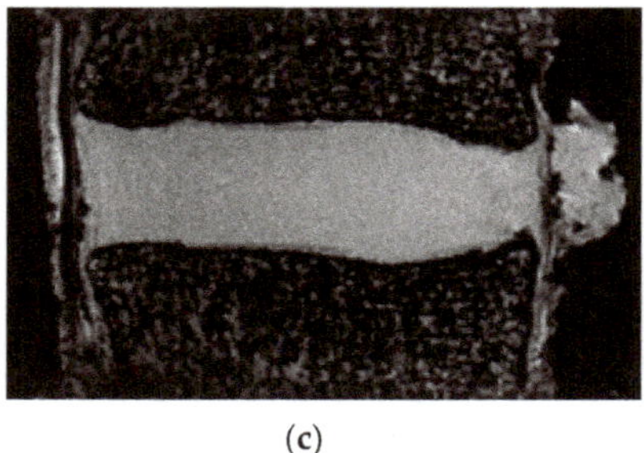
(c)

Figure 1. Provoked herniation with clear nucleus extrusion with (**a**) endoscopic view of the posterior annulus fibrosus and imaging from ultra-high field MRI (11.7 T) (**b**) in the transverse and (**c**) sagittal plane.

3.1.1. Influence of Shape and Size of the Annular Defect

After setting the first vertical defect, a herniation could only be provoked in one specimen. Enlarging the defect to a size of 1.0 mm × 5.5 mm led to a herniation in one other specimen. In three other specimens, a herniation could be provoked after the annular defect had a size of 1.2 mm × 6.5 mm. One of those specimens showed an intermittent herniation with nucleus pulposus material protruding through the annulus fibrosus in the phase of load application and wandering back into the interior of the disc when unloading. In one specimen, no herniation could be provoked at all. In the specimens with round annular defects, a herniation could be provoked in six specimens after setting a defect with a diameter of Ø 4 mm. In two specimens, no herniation could be provoked at all, neither after widening the defect diameter to Ø 6 mm nor to Ø 8 mm.

3.1.2. Influence of Daily-Life Activities

In seven of the 11 herniated discs (Table 3), the activity *lifting boxes* led to a herniation, whereby four herniations occurred under pure flexion and three herniations occurred under complex motion. One herniation could be provoked while simulating *sweeping floor*. Three discs with a round defect (Ø 4 mm) already herniated during the flexibility test before dynamic loading.

3.2. Ultra-High Field MR Imaging

The herniations and defects of the specimens with vertical defects could be observed and further investigated through highly resolved images of the ultra-high field MRI (Figure 1b,c). It could be observed that the herniated nucleus pulposus material always extruded through the defect that was previously set. The sequester extruded to the posterocentral or posterolateral side of the disc.

3.3. Biomechanical Parameters

The herniation itself and the defect only led to a slight increase in ROM and NZ by overall about 1° (Figure 2), and a very slight decrease of IDP (Figure 3). After setting the vertical defects, IDP slightly decreased by 0.2 MPa and 0.1 MPa after setting the round defects. It could be observed that stronger disc injury resulted in a wider range especially for NZ, but also for ROM and IDP values. However, no significant differences could be observed either between specimens with vertical and round shaped defect or between the different conditions of the specimens.

Interestingly, the specimens that did not herniate at all showed slightly lower IDP values or higher ROM either already in the intact state or after creation of the defect.

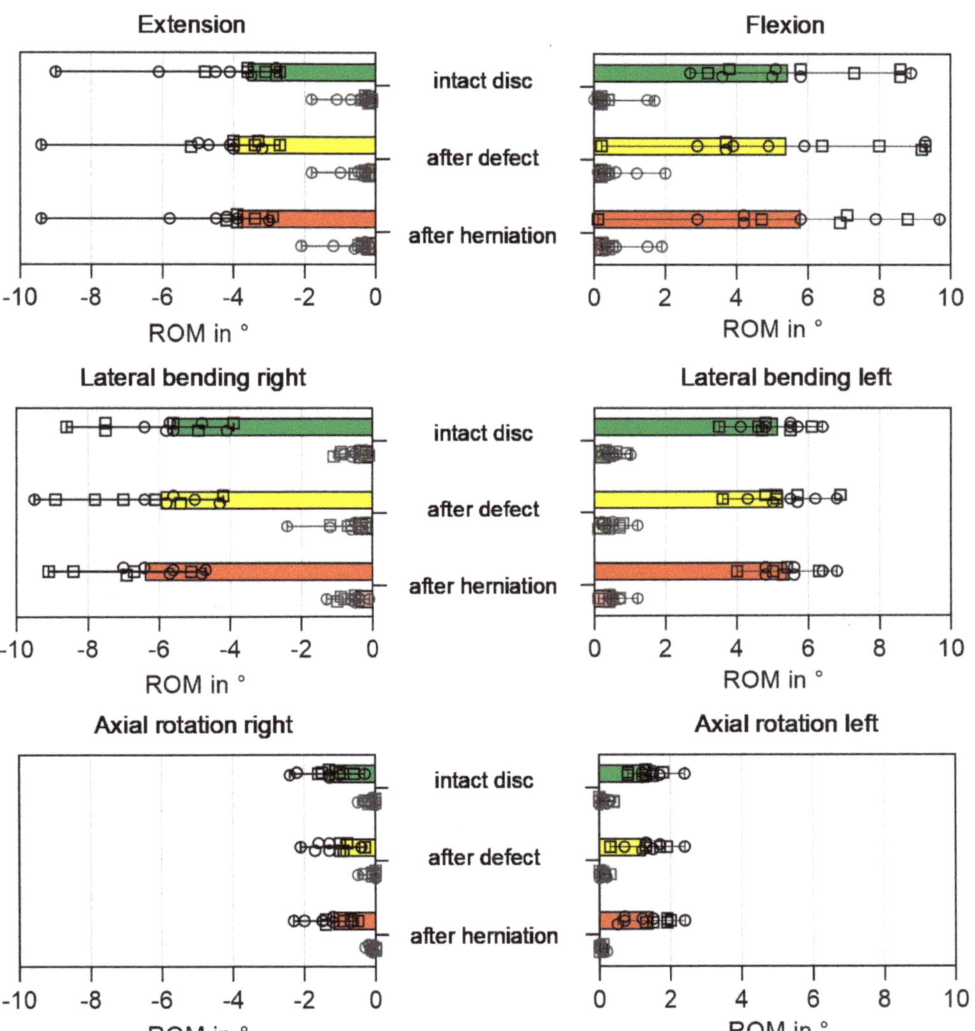

Figure 2. Range of motion (ROM) in ° in flexion-extension, left/right lateral bending and left/right axial rotation for the intact disc, after setting the vertical (□) or round (o) defects, respectively, assessed with the universal spine tester under pure moments of ±7.5 Nm.

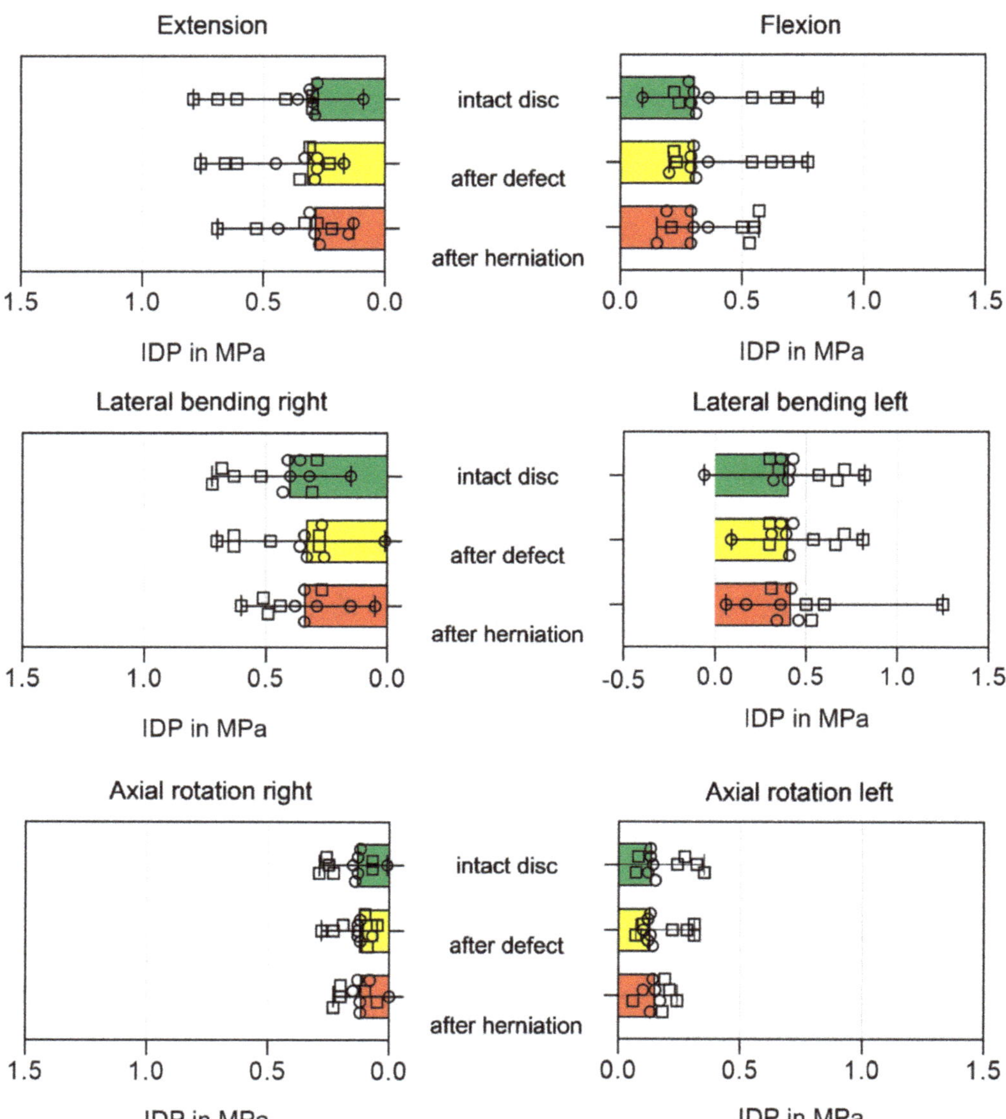

Figure 3. Intradiscal pressure (IDP) in MPa in flexion-extension, left/right lateral bending and left/right axial rotation for the intact disc, after setting the vertical (□) or round (o) defects, respectively, assessed with the universal spine tester under pure moments of ±7.5 Nm.

4. Discussion

In this study, a new physiological test method was developed that allows the replication of different physiological activities in vitro. With this dynamical test method, it was possible to provoke lumbar disc herniations under experimental conditions. A lumbar disc herniation model was developed that can be used for the biomechanical investigation of new regenerative therapies in order to prevent or treat intervertebral disc herniations.

The simulation of different physiological daily-life activities of patients resulted in intravertebral disc herniations, but only after an annular defect was previously set. The annular defects mimicked a structural failure of the annulus fibrosus which has also been observed in patients with herniated intervertebral discs through clinical MRI. Moreover, annular tears have also been identified in asymptomatic individuals. It could be concluded, that an intervertebral disc herniation with nucleus extrusion through an annular defect might be the consequence of substantial annular fissures or tears [28].

However, a herniation could not be provoked in every specimen, even with a large annular defect. The observation of this specimen with ultra-high field MRI might indicate that structural changes have already occurred inside the disc. Compared to the other intervertebral discs, the nucleus pulposus did not look as homogeneous as in the herniated discs and showed disturbances in the signal response of the MRI (Figure 4). It can be assumed that the ability of the disc to create hydrostatic pressure was already slightly impaired. This might underlie the assumption of Wilke et al. that the risk of getting a herniation is higher in younger patients with non-degenerated discs [42], but only if annular defects coexist. This goes in accordance with the findings of Adams et al. that the prevalence of disc herniations increases with age because the annulus fibrosus or endplate junction becomes increasingly injured over time [2]. Additionally, the findings of this study confirm the hypothesis that the risk of an intervertebral disc herniation might be generally higher in the morning after the intervertebral disc has been rehydrated by recovering during night rest [43].

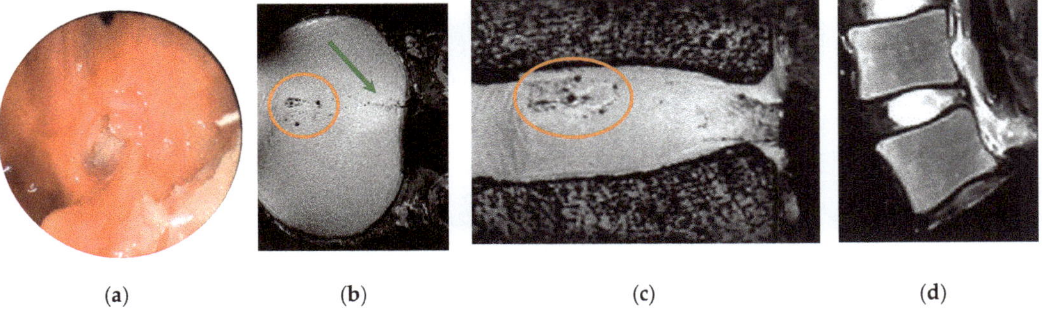

(a) (b) (c) (d)

Figure 4. Specimen in which no herniation could be provoked (**a**) endoscopic view of the posterior annulus fibrosus and imaging from ultra-high field MRI (11.7 T) (**b**) in the transverse and (**c**) sagittal plane cutting through the defect that can be clearly seen, indicated with a green arrow in (**b**). In both ultra-high field MRI sections (**b**,**c**), disturbances of the nucleus pulposus structure and maybe air inclusions could be detected (orange circles). Those disturbances could not be detected in the (**d**) MRI scans performed prior to testing.

In all specimens, the extrusion of nucleus pulposus occurred either to the central region or to the opposite site to where the combined motion was exerted. This behavior was already observed by Adams and Hutton [2].

Three discs with a round annular defect were already herniated during the flexibility test before dynamic loading. During the flexibility test, the IDP values created by pure motion, were comparable to the in vivo IDP in a standing position. A reason for the early occurrence of a nucleus extrusion in the discs with round annular defects could be that the initial damage led to a larger expansion of the defect due to high fiber strains generated by the dynamical loading [44], compared to the thinner initial vertical defect. Hence for further experiments, we would suggest using a model with a round defect in order to guarantee the successful provocation of a nucleus extrusion.

The defect and the extrusion of nucleus pulposus material through the aforesaid led to a migration of nucleus material. Interestingly, this migration only caused a small decrease of IDP and increase of ROM. From former in vitro studies [42,45] it was known, that treating a disc herniation by partial nucleotomy significantly decreased the IDP and increased ROM. From these findings, it can be concluded that only changes in nucleus volume lead to a significant change of the biomechanical properties, whereas volume migrations do not.

In this study, only herniations with clear nucleus pulposus extrusion through an annular defect could be provoked and hence, investigated. So far, no intervertebral disc herniations caused by endplate junction failures [16] could be investigated by using this test method and it cannot be answered yet as to which activities might lead to endplate junction failures. However, using this new test method, other activities could be simulated in vitro and crucial motion of loading scenarios could be identified that might lead to endplate junction failures.

Furthermore, the defects were set artificially. In previous in vitro studies, it could be shown that annular defects are usually initiated from the inside of the disc and migrate to the outer layers of the annulus fibrosus [26]. Such annular defects were provoked during long-term excessive dynamical loading or under exaggerated loading conditions [2,15,26]. Long-term and excessive dynamical test protocols result in a completely dehydrated disc and especially nucleus pulposus which is unable to rehydrate completely under experimental conditions [34]. Hence, after previously excessive loading of the specimens, it would not have been possible to generate hydrostatic pressure conditions inside the discs [22], which was the basis for this new test method.

Another limitation of this study is that no in vivo IDP values were available for the activities with complex, combined motion. It was assumed that the IDP does change, but the axial load does not change between corresponding activities with single flexion or combined flexion, lateral bending, and axial rotation. The results of the validation might confirm this approach as hydrostatic IDP values highly depend on the individual disc quality. Nevertheless, it would be beneficial to investigate in greater detail how external loads and muscle forces are distributed in the intervertebral disc.

5. Conclusions

In this study, a new lumbar disc herniation model was established by which regenerative tissue repair approaches as well as novel implants, such as nucleus implants or annulus sealing methods for the treatment of lumbar disc herniations could be challenged under normal and physiological worst-case scenarios. Therefore, a new biomechanical in vitro test method was developed that dynamically simulates daily-life activities. By means of this test method, disc herniations could be successfully produced and the influence of shape and size of annular defects investigated. Based on these results, it is recommended to produce a disc herniation in non- or only low degenerated human lumbar discs through a round annular defect.

Supplementary Materials: The following are available online at https://www.mdpi.com/2076-3417/11/6/2847/s1, Table S1: Ultra-high field MRI-IVDs, Video S1: Dynamic disc loading simulator with simulation of physiological activities, Video S2: Lumbar Disc Herniation.

Author Contributions: Conceptualization, H.-J.W. and L.Z. (Laura Zengerle); methodology, L.Z. (Laura Zengerle) and E.D.; software, L.Z. (Laura Zengerle) and E.D.; validation, E.D. and L.Z. (Laura Zengerle); formal analysis, E.D., L.Z. (Laura Zengerle), B.K. and L.Z. (Lena Zöllner); investigation, L.Z. (Laura Zengerle), E.D. and L.Z. (Lena Zöllner); resources, H.-J.W.; data curation, E.D., L.Z. (Lena Zöllner); writing—original draft preparation, L.Z. (Laura Zengerle); writing—review and editing, H.-J.W., E.D., B.K., L.Z. (Lena Zöllner); visualization, B.K., E.D., L.Z. (Laura Zengerle), L.Z. (Lena Zöllner); supervision, H.-J.W., L.Z. (Laura Zengerle); project administration, H.-J.W.; funding acquisition, H.-J.W. All authors have read and agreed to the published version of the manuscript.

Funding: This research was funded by the German Research Foundation (DFG), grant number Wi1352/14-3.

Institutional Review Board Statement: The study was conducted according to the guidelines of the Declaration of Helsinki, and approved by the Institutional Review Board (or Ethics Committee) of Ulm University (protocol code 35/16 and date of approval 3 October 2016).

Informed Consent Statement: Not applicable.

Data Availability Statement: The data presented in this study are available on request from the corresponding author. The data are not publicly available due to confidentiality reasons.

Acknowledgments: The authors thank the Core Facility Small Animal Imaging and the Ulm University Center for Translational Imaging MoMAN for their support. The authors further acknowledge Youping Tao, Jan Ulrich Jansen, Theresa Schilpp and Saskia Brendle for their help in the experiments.

Conflicts of Interest: The authors declare no conflict of interest.

References

1. Evans, W.; Jobe, W.; Seibert, C. A cross-sectional prevalence study of lumbar disc degeneration in a working population. *Spine* **1989**, *14*, 60–64. [CrossRef]
2. Adams, M.A.; Hutton, W.C. The mechanics of prolapsed intervertebral disc. *Int. Orthop.* **1982**, *6*, 249–253. [CrossRef]
3. Spangfort, E.V. The lumbar disc herniation. A computer-aided analysis of 2504 operations. *Acta Orthop. Scand.* **1972**, *142*, 1–95. [CrossRef]
4. White, A.A.; Panjabi, M.M.; Company, J.B.L. *Clinical Biomechanics of the Spine*; J. B. Lippincott Company: Philadelphia, PA, USA, 1978.
5. Iatridis, J.C.; Nicoll, S.B.; Michalek, A.J.; Walter, B.A.; Gupta, M.S. Role of biomechanics in intervertebral disc degeneration and regenerative therapies: What needs repairing in the disc and what are promising biomaterials for its repair? *Spine J.* **2013**, *13*, 243–262. [CrossRef] [PubMed]
6. Sloan, S.R., Jr.; Wipplinger, C.; Kirnaz, S.; Navarro-Ramirez, R.; Schmidt, F.; McCloskey, D.; Pannellini, T.; Schiavinato, A.; Härtl, R.; Bonassar, L.J. Combined nucleus pulposus augmentation and annulus fibrosus repair prevents acute intervertebral disc degeneration after discectomy. *Sci. Transl. Med.* **2020**, *12*. [CrossRef] [PubMed]
7. Zengerle, L.; Köhler, A.; Debout, E.; Hackenbroch, C.; Wilke, H.-J. Nucleus replacement could get a new chance with annulus closure. *Eur. Spine J.* **2020**, *29*, 1733–1741. [CrossRef] [PubMed]
8. Bowles, R.D.; Setton, L.A. Biomaterials for intervertebral disc regeneration and repair. *Biomaterials* **2017**, *129*, 54–67. [CrossRef]
9. Huang, Y.C.; Hu, Y.; Li, Z.; Luk, K.D.K. Biomaterials for intervertebral disc regeneration: Current status and looming challenges. *J. Tissue Eng. Regen. Med.* **2018**, *12*, 2188–2202. [CrossRef]
10. Huang, Y.C.; Urban, J.P.; Luk, K.D. Intervertebral disc regeneration: Do nutrients lead the way? *Nat. Rev. Rheumatol.* **2014**, *10*, 561–566. [CrossRef] [PubMed]
11. Sakai, D.; Andersson, G.B. Stem cell therapy for intervertebral disc regeneration: Obstacles and solutions. *Nat. Rev. Rheumatol.* **2015**, *11*, 243–256. [CrossRef]
12. Campisi, M.; Shin, Y.; Osaki, T.; Hajal, C.; Chiono, V.; Kamm, R.D. 3D self-organized microvascular model of the human blood-brain barrier with endothelial cells, pericytes and astrocytes. *Biomaterials* **2018**, *180*, 117–129. [CrossRef]
13. Foresti, R.; Rossi, S.; Pinelli, S.; Alinovi, R.; Barozzi, M.; Sciancalepore, C.; Galetti, M.; Caffarra, C.; Lagonegro, P.; Scavia, G. Highly-defined bioprinting of long-term vascularized scaffolds with Bio-Trap: Complex geometry functionalization and process parameters with computer aided tissue engineering. *Materialia* **2020**, *9*, 100560. [CrossRef]
14. Foresti, R.; Rossi, S.; Pinelli, S.; Alinovi, R.; Sciancalepore, C.; Delmonte, N.; Selleri, S.; Caffarra, C.; Raposio, E.; Macaluso, G.; et al. In-vivo vascular application via ultra-fast bioprinting for future 5D personalised nanomedicine. *Sci. Rep.* **2020**, *10*, 3205. [CrossRef] [PubMed]
15. Berger-Roscher, N.; Casaroli, G.; Rasche, V.; Villa, T.; Galbusera, F.; Wilke, H.J. Influence of Complex Loading Conditions on Intervertebral Disc Failure. *Spine* **2017**, *42*, E78–E85. [CrossRef]
16. Rajasekaran, S.; Bajaj, N.; Tubaki, V.; Kanna, R.M.; Shetty, A.P. ISSLS Prize winner: The anatomy of failure in lumbar disc herniation: An in vivo, multimodal, prospective study of 181 subjects. *Spine* **2013**, *38*, 1491–1500. [CrossRef] [PubMed]
17. Ahsan, M.K.; Matin, T.; Ali, M.I.; Ali, M.Y.; Awwal, M.A.; Sakeb, N. Relationship between physical work load and lumbar disc herniation. *Mymensingh Med. J.* **2013**, *22*, 533–540. [PubMed]
18. Kelsey, J.L.; Githens, P.B.; Walter, S.D.; Southwick, W.O.; Weil, U.; Holford, T.R.; Ostfeld, A.M.; Calogero, J.A.; O'Connor, T.; White, A.A. An epidemiological study of acute prolapsed cervical intervertebral disc. *J. Bone Joint Surg. Am.* **1984**, *66*, 907–914. [CrossRef] [PubMed]
19. Mundt, D.J.; Kelsey, J.L.; Golden, A.L.; Panjabi, M.M.; Pastides, H.; Berg, A.T.; Sklar, J.; Hosea, T. An epidemiologic study of sports and weight lifting as possible risk factors for herniated lumbar and cervical discs. The Northeast Collaborative Group on Low Back Pain. *Am. J. Sports Med.* **1993**, *21*, 854–860. [CrossRef]
20. Pietila, T.A.; Stendel, R.; Kombos, T.; Ramsbacher, J.; Schulte, T.; Brock, M. Lumbar disc herniation in patients up to 25 years of age. *Neurol. Med. Chir.* **2001**, *41*, 340–344. [CrossRef]
21. Hutton, W.C.; Adams, M.A. Can the lumbar spine be crushed in heavy lifting? *Spine* **1982**, *7*, 586–590.

22. Adams, M.A.; Freeman, B.J.; Morrison, H.P.; Nelson, I.W.; Dolan, P. Mechanical initiation of intervertebral disc degeneration. *Spine* **2000**, *25*, 1625–1636. [CrossRef]
23. Gordon, S.J.; Yang, K.H.; Mayer, P.J.; Mace, A.H., Jr.; Kish, V.L.; Radin, E.L. Mechanism of disc rupture. A preliminary report. *Spine* **1991**, *16*, 450–456. [CrossRef]
24. Lu, Y.M.; Hutton, W.C.; Gharpuray, V.M. Do bending, twisting, and diurnal fluid changes in the disc affect the propensity to prolapse? A viscoelastic finite element model. *Spine* **1996**, *21*, 2570–2579. [CrossRef] [PubMed]
25. McNally, D.S.; Adams, M.A.; Goodship, A.E. Can intervertebral disc prolapse be predicted by disc mechanics? *Spine* **1993**, *18*, 1525–1530. [CrossRef] [PubMed]
26. Wade, K.R.; Robertson, P.A.; Thambyah, A.; Broom, N.D. How healthy discs herniate: A biomechanical and microstructural study investigating the combined effects of compression rate and flexion. *Spine* **2014**, *39*, 1018–1028. [CrossRef]
27. Wade, K.R.; Robertson, P.A.; Thambyah, A.; Broom, N.D. "Surprise" Loading in Flexion Increases the Risk of Disc Herniation Due to Annulus-Endplate Junction Failure: A Mechanical and Microstructural Investigation. *Spine* **2015**, *40*, 891–901. [CrossRef]
28. Wade, K.R.; Schollum, M.L.; Robertson, P.A.; Thambyah, A.; Broom, N.D. ISSLS Prize Winner: Vibration Really Does Disrupt the Disc: A Microanatomical Investigation. *Spine* **2016**, *41*, 1185–1198. [CrossRef] [PubMed]
29. Edwards, W.T.; Ordway, N.R.; Zheng, Y.; McCullen, G.; Han, Z.; Yuan, H.A. Peak stresses observed in the posterior lateral anulus. *Spine* **2001**, *26*, 1753–1759. [CrossRef]
30. Wade, K.; Berger-Roscher, N.; Rasche, V.; Wilke, H. Disc wall structural abnormalities can act as initiation sites for herniation. *Eur. Cell Mater.* **2020**, *40*, 227–238. [CrossRef]
31. Behjati, M.; Arjmand, N. Biomechanical Assessment of the NIOSH Lifting Equation in Asymmetric Load-Handling Activities Using a Detailed Musculoskeletal Model. *Hum. Factors* **2018**, *61*, 191–202. [CrossRef] [PubMed]
32. Gunzburg, R.; Parkinson, R.; Moore, R.; Cantraine, F.; Hutton, W.; Vernon-Roberts, B.; Fraser, R. A cadaveric study comparing discography, magnetic resonance imaging, histology, and mechanical behavior of the human lumbar disc. *Spine* **1992**, *17*, 417–426. [CrossRef]
33. Andersson, G.B. Epidemiologic aspects on low-back pain in industry. *Spine* **1981**, *6*, 53–60. [CrossRef]
34. Pflaster, D.S.; Krag, M.H.; Johnson, C.C.; Haugh, L.D.; Pope, M.H. Effect of test environment on intervertebral disc hydration. *Spine* **1997**, *22*, 133–139. [CrossRef]
35. Wilke, H.J.; Jungkunz, B.; Wenger, K.; Claes, L.E. Spinal segment range of motion as a function of in vitro test conditions: Effects of exposure period, accumulated cycles, angular-deformation rate, and moisture condition. *Anat. Rec.* **1998**, *251*, 15–19. [CrossRef]
36. Pfirrmann, C.W.; Metzdorf, A.; Zanetti, M.; Hodler, J.; Boos, N. Magnetic resonance classification of lumbar intervertebral disc degeneration. *Spine* **2001**, *26*, 1873–1878. [CrossRef] [PubMed]
37. Pearcy, M.J. Stereo radiography of lumbar spine motion. *Acta Orthop. Scand.* **1985**, *212*, 1–45. [CrossRef]
38. Wilke, H.J.; Neef, P.; Caimi, M.; Hoogland, T.; Claes, L.E. New in vivo measurements of pressures in the intervertebral disc in daily life. *Spine* **1999**, *24*, 755–762. [CrossRef]
39. Wilke, H.J.; Kienle, A.; Maile, S.; Rasche, V.; Berger-Roscher, N. A new dynamic six degrees of freedom disc-loading simulator allows to provoke disc damage and herniation. *Eur. Spine J.* **2016**, *25*, 1363–1372. [CrossRef] [PubMed]
40. Wilke, H.J.; Claes, L.; Schmitt, H.; Wolf, S. A universal spine tester for in vitro experiments with muscle force simulation. *Eur. Spine J.* **1994**, *3*, 91–97. [CrossRef]
41. Wilke, H.J.; Wenger, K.; Claes, L. Testing criteria for spinal implants: Recommendations for the standardization of in vitro stability testing of spinal implants. *Eur. Spine J.* **1998**, *7*, 148–154. [CrossRef]
42. Wilke, H.J.; Ressel, L.; Heuer, F.; Graf, N.; Rath, S. Can prevention of a reherniation be investigated? Establishment of a herniation model and experiments with an anular closure device. *Spine* **2013**, *38*, E587–E593. [CrossRef] [PubMed]
43. Belavy, D.L.; Adams, M.; Brisby, H.; Cagnie, B.; Danneels, L.; Fairbank, J.; Hargens, A.R.; Judex, S.; Scheuring, R.A.; Sovelius, R.; et al. Disc herniations in astronauts: What causes them, and what does it tell us about herniation on earth? *Eur. Spine J.* **2016**, *25*, 144–154. [CrossRef] [PubMed]
44. Schmidt, H.; Kettler, A.; Heuer, F.; Simon, U.; Claes, L.; Wilke, H.J. Intradiscal pressure, shear strain, and fiber strain in the intervertebral disc under combined loading. *Spine* **2007**, *32*, 748–755. [CrossRef] [PubMed]
45. Heuer, F.; Ulrich, S.; Claes, L.; Wilke, H.J. Biomechanical evaluation of conventional anulus fibrosus closure methods required for nucleus replacement. Laboratory investigation. *J. Neurosurg. Spine* **2008**, *9*, 307–313. [CrossRef] [PubMed]

Article

Synergistic Effects of Acidic pH and Pro-Inflammatory Cytokines IL-1β and TNF-α for Cell-Based Intervertebral Disc Regeneration

Chiara Borrelli [1,2] and Conor T. Buckley [1,2,3,4,*]

1 Trinity Centre for Biomedical Engineering, Trinity Biomedical Sciences Institute, Trinity College Dublin, The University of Dublin, D02 R590 Dublin, Ireland; borrellc@tcd.ie
2 Discipline of Mechanical, Manufacturing and Biomedical Engineering, School of Engineering, Trinity College Dublin, The University of Dublin, D02 PN40 Dublin, Ireland
3 Advanced Materials and Bioengineering Research (AMBER) Centre, Royal College of Surgeons in Ireland & Trinity College Dublin, The University of Dublin, D02 PN40 Dublin, Ireland
4 Tissue Engineering Research Group, Department of Anatomy and Regenerative Medicine, Royal College of Surgeons in Ireland, 121/122 St. Stephen's Green, D02 H903 Dublin, Ireland
* Correspondence: conor.buckley@tcd.ie; Tel.: +353-1-896-2061

Received: 26 November 2020; Accepted: 14 December 2020; Published: 17 December 2020

Abstract: The intervertebral disc (IVD) relies mainly on diffusion through the cartilaginous endplates (CEP) to regulate the nutrient and metabolites exchange, thus creating a challenging microenvironment. Degeneration of the IVD is associated with intradiscal acidification and elevated levels of pro-inflammatory cytokines. However, the synergistic impact of these microenvironmental factors for cell-based therapies remains to be elucidated. The aim of this study was to investigate the effects of low pH and physiological levels of interleukin-1ß (IL-1β) and tumour necrosis factor-α (TNF-α) on nasal chondrocytes (NCs) and subsequently compare their matrix forming capacity to nucleus pulposus (NP) cells in acidic and inflamed culture conditions. NCs and NP cells were cultured in low glucose and low oxygen at different pH conditions (pH 7.1, 6.8 and 6.5) and supplemented with physiological levels of IL-1β and TNF-α. Results showed that acidosis played a pivotal role in influencing cell viability and matrix accumulation, while inflammatory cytokine supplementation had a minor impact. This study demonstrates that intradiscal pH is a dominant factor in determining cell viability and subsequent cell function when compared to physiologically relevant inflammatory conditions. Moreover, we found that NCs allowed for improved cell viability and more effective NP-like matrix synthesis compared to NP cells, and therefore may represent an alternative and appropriate cell choice for disc regeneration.

Keywords: nasal chondrocytes; disc degeneration; inflammation; microenvironment; nucleus pulposus; cell; spine

1. Introduction

Low back pain (LBP) is a leading cause of disability, affecting more than 600 million people worldwide [1]. Although it is well established that the causes of LBP are multifactorial and can include mechanical injury [2], genetic predisposition [3] or even lifestyle activities [4], a significant proportion of the cases are associated with intervertebral disc (IVD) degeneration [5,6]. The IVD is an avascular organ that interfaces the superior and inferior vertebral bodies via the cartilaginous endplates (CEPs), regulating the nutrient and waste metabolites path into and out of the IVD, respectively [7]. The microenvironment of a degenerative IVD is considered a hostile niche characterised by large concentration gradients of nutrients and metabolites across its domain, with significantly lower glucose and oxygen concentrations in the centre of the disc, the nucleus pulposus (NP), compared to the periphery, the annulus fibrosus (AF) [8]. Due to the low levels of oxygen in the centre of the IVD, cells residing in the NP rely mainly on anaerobic respiration to create energy, producing lactate as a by-product of glycolysis, and thereby causing acidification of the local microenvironment. With the onset of disc degeneration CEPs undergo physical changes such as thinning and calcification [9,10], thereby reducing the bidirectional flow of nutrients and metabolites to and from the NP region causing local lactate accumulation and a consequent decrease in pH [11]. It has been shown that the pH of the disc changes with the degree of degeneration, ranging from pH 7.1 to pH 6.5 in healthy to severely degenerated conditions respectively [12,13] and that the decrease in the pH in the organ is directly correlated with matrix catabolism [14,15] and reduced cell viability [16].

The exact mechanisms that trigger degeneration of the IVD are unknown, although it is believed to be mediated by the abnormal production of pro-inflammatory cytokines by NP cells [17,18]. These cytokines are believed to trigger a series of cellular responses that promote cell senescence, autophagy and apoptosis [19–21]. While a number of pro-inflammatory cytokines have been identified in degenerated discs, tumour necrosis factor-α (TNF-α) and interleukin-1β (IL-1β) are believed to play central roles [21–23]. Both IL-1β and TNF-α have been identified to being involved in disc herniation [24], nerve ingrowth [25] and in the upregulation of genes encoding matrix-degrading enzymes [26–28]. Although the correlation of IL-1β and TNF-α with disc degeneration is well established in vivo, the response to their in vitro supplementation in 2D and 3D cultures has been diverse and often conflicting [19,23]. One explanation for the differential response observed from in vitro experimentation could be the concentration of supplemented cytokines being supraphysiological. In a recent study, Altun et al. measured the concentration of IL-1β and TNF-α in degenerated human discs and found their levels to be notably increased in patients affected by acute IVD degeneration, with concentrations in the order of pg/mL [29]. Nevertheless, the concentration of inflammatory cytokines used in the majority of experiments investigating their effects on cells in vitro is usually found to be in the order of ng/mL, and significantly higher compared to in vivo [30–32].

The role of inflammatory cytokines and environmental pH in IVD degeneration can be linked to many detrimental effects, however it is difficult to determine which of the two has a primary role in disc disease. To fully appreciate the effectiveness of any proposed cell therapy for disc regeneration for clinical use, the regenerative capacity of the cellular component must be examined under physiologically relevant culture conditions.

Identifying an appropriate cell source for disc repair has received significant attention over the last decade, and in particular, the attractiveness of using a patient's own cellular material to avoid complications in relation with immune rejection, supply, ethical and regulatory considerations has been highlighted [33,34]. Among the various autologous cell sources evaluated in the literature, nasal chondrocytes (NCs) have recently been explored as a potential autologous cell source for cartilage repair [35–37] and may represent a valid alternative for IVD repair strategies [38]. We have previously demonstrated the ability of NCs to remain viable and functional in response to NP-like oxygen and glucose levels [38]. However, the ability of NCs to maintain the same viability and functionality in an inflamed and acidic microenvironment remains to be elucidated.

Hence, the primary objective of this study was to determine whether physiologically relevant concentrations of pro-inflammatory cytokines IL-1β and TNF-α, or an acidic microenvironment impacts cell survival and matrix production of NCs cultured in low glucose and low oxygen conditions. Finally, we compared the functional matrix synthesis capacity of NCs and NP cells in representative physiological conditions of degeneration.

2. Materials and Methods

2.1. Cell Isolation and Monolayer Expansion

Porcine nasal tissue was sourced from a local abattoir and dissected within 24 h. NC isolation was performed as previously described [39]. Briefly, cartilage from the nasal septum was minced

(~2 mm) and digested in serum free Low Glucose–Dulbecco's Modified Eagle Medium (LG-DMEM) containing penicillin (100 U/mL)–streptomycin (100 μg/mL) (PenStrep) and 3000 U/mL of collagenase type II (Gibco, Invitrogen, Dublin, Ireland) at a ratio of 10 mL per gram of minced tissue. The digestion of minced tissue was performed under constant rotation for 3 h at 37 °C and subjected to physical agitation using a tissue dissociator (gentleMACS™, Miltenyi Biotech, Surrey, UK). Cells were separated from tissue residues (40 μm cell strainer) and trypan blue exclusion was used to determine cell yield and viability. NCs were seeded in T-175 flasks (5×10^3 cells/cm^2) and expanded to passage two (P2).

NP cells were isolated from the IVDs of porcine spines. Briefly, NP tissues were harvested aseptically and minced. Tissue fragments were placed in T-25 flasks containing LG-DMEM with 10% FBS and PenStrep and cultured in a humidified atmosphere at 37 °C and 5% O_2. Once cell migration from the tissue had occurred, flasks were washed to remove debris and NP cells were expanded to 80% confluence and transferred to T-175 flasks (5×10^3 cells/cm^2) and expanded to passage two (P2). All expansion cultures contained LG-DMEM supplemented with 10% FBS, 2% PenStrep and were maintained at 37 °C and 5% O_2.

2.2. Preparation of Media for Experimental Culture

All media formulations were prepared from Chemically Defined Medium (CDM) containing LG-DMEM supplemented with 0.25 μg/mL amphotericin B, 2% PenStrep, 100 nM Dexamethasone, 50 μg/mL L-ascorbic acid-2-phosphate, 1% insulin-transferrin-selenium, 4.7 μg/mL linoleic acid, 40 μg/mL L-proline and 1.5 mg/mL bovine serum albumin (BSA). Media was adjusted to pH 7.1, 6.8 and 6.5 by the addition of 400 μL, 450 μL and 500 μL of 3M HCl, respectively, and 40 μL of 5M lactic acid (LacA) to 50 mL of CDM to obtain physiological lactate levels (4 mM) normally found in the IVD [11]. Acidic media was subsequently incubated overnight in a humidified atmosphere at 37 °C and 5% O_2.

2.3. 2D Culture under Varying pH and Inflammatory Conditions

Expanded NCs were trypsinised, counted using trypan blue staining and seeded into T-25 flasks at a density of 1×10^4 cells/cm^2. Cells were allowed to adhere to the culture plastic overnight, and media was changed to pH modified CDM (pH 7.1, 6.8 and 6.5) containing no added cytokines (Control), 125 pg/mL IL-1β, 25 pg/mL TNF-α, or 125 pg/mL IL-1β and 25 pg/mL TNF-α for 7 days at 37 °C and 5% O_2. One complete media exchange was performed four days after seeding. 2D culture effects were assessed in terms of cell counts with trypan blue stain at days 0 and 7, and cell density/morphology was visualised using crystal violet staining.

2.4. Pre-Gel Fabrication, Cell Encapsulation and Culture

Cells were encapsulated in 3D hydrogels fabricated using a disc extracellular matrix (ECM) derived biomaterial previously developed in our laboratory [40]. Briefly, disc ECM was prepared by solubilizing cryomilled powder in 0.5 M acetic acid (4% *w/v*) containing pepsin (2.5 mg/mL, Sigma). Solubilised ECM (sECM) and *N*-hydroxysuccinimide (NHS) functionalised CS (fCS) were combined to yield a final gel composition of 2% sECM-2% fCS. 5% *v/v* 10x phosphate buffered saline (PBS) was added to the solution and the pH was adjusted to 7.4. A suspension of NCs or NP cells was added to the pre-gel at a cell density of 1×10^6 cells/mL. Cell seeded pre-gel was cast into cylindrical moulds of 6 mm diameter and incubated at 37 °C for 1 h. Following gelation, the hydrogel constructs were cultured in pH modified CDM (pH 7.1, 6.8 and 6.5) containing no added cytokines (Control), 125 pg/mL IL-1β or 25 pg/mL TNF-α, or a combination of 125 pg/mL IL-1β and 25 pg/mL TNF-α, and incubated for 14 days at 37 °C and 5% O_2. In the second part of the study pH 6.8 groups and groups containing IL-1β or TNF-α only were excluded. Media was changed twice weekly and the supernatant was stored at 4 °C for biochemical analysis.

2.5. DNA, Sulphated Glycosaminoglycan and Collagen Content

On termination of culture, samples were stored at −80 °C until further analysis. Digestion of samples was performed under constant agitation (60 °C, 12 h) with 100 mM sodium phosphate/5mM Na_2EDTA (pH 6.5) and papain enzyme (3.88 U/mL) containing L-cysteine (5 mM). Hoechst Bisbenzimide 33258 dye (DNA QF Kit, Sigma-Aldrich, Arklow, Ireland) was used for DNA quantification, while the dimethylene blue dye binding assay was used to determine sulphated glycosaminoglycan (sGAG) content. For determining hydroxyproline content, samples were hydrolysed (110 °C, 18 h) with HCL (38%) and assayed using chloramine-T [41], and the collagen content determined using a hydroxyproline:collagen ratio of 1:7.69 [42]. Samples of media supernatants were also analysed for both sGAG and collagen content.

2.6. Assessment and Analysis of Cell Viability

A LIVE/DEAD® (Invitrogen, Dublin, Ireland) assay was used to determine cell viability. Following a PBS washing step, samples were incubated for 1 h in phenol free LG-DMEM containing Calcein AM (4 µM) and Ethidium Homodimer 1 (4 µM) (Cambridge Bioscience, Bar Hill, UK). Images were captured using an Olympus FV-1000 Point-Scanning Confocal Microscope (515 nm, 615 nm wavelengths). Semi-quantitative analysis of cell viability was determined using ImageJ software (ImageJ, NIH, Bethesda, MD, USA).

2.7. Histology and Immunohistochemistry

Samples were washed in PBS, treated with 4% paraformaldehyde (4 °C, 12 h), dehydrated in a series of graded alcohols and finally wax embedded. Sections of 8 µm were stained with 1% Alcian Blue 8GX in 0.1 M HCl to assess sGAG deposition. Collagen types I and II were assessed using immunohistochemistry techniques. Sections were treated with chondroitinase ABC (37 °C, 1 h) (Sigma-Aldrich), and non-specific sites were blocked using 5% BSA. Collagen type I (Abcam, Cambridge, UK) and collagen type II (Santa Cruz, Heidelberg, Germany) primary antibodies incubated at 4 °C overnight. The secondary antibody (Anti-Mouse igG biotin conjugate, Sigma-Aldrich) was applied for 1 h followed by incubation (45 min) with ABC reagent (Vectastain PK-400, Vector Labs, Upper Heyford, UK). DAB peroxidase (Vector Labs, UK) was used as a developer.

2.8. Statistical Analysis

GraphPad Prism (Ver. 7) was used for presentation of graphical data (mean ± standard deviation) and statistical analysis (two-way ANOVA), with significance accepted at a level of $p < 0.05$. N is used to represent the number of biological donors and n to represent the technical replicates for each experiment performed.

3. Results

3.1. pH and Inflammatory Cytokine Effects on NC in 2D Culture

There was an increase in the number of viable cells after 7 days of 2D culture for pH 7.1 control and pH 7.1 IL-1β groups compared to day 0 levels. Conversely, TNF-α and IL-1β+TNF-α groups did not exhibit significant changes compared to day 0 samples in terms of cell number (Figure 1A). As expected, there was a decrease in the number of viable cells with increasing acidity. However, it is unclear whether lower cell numbers were as a result of diminished proliferation, increased cell death or a combination of both. Cellular density as assessed through crystal violet staining confirmed the biochemical results with no noticeable differences in cell morphology observed for the different groups investigated (Figure 1B).

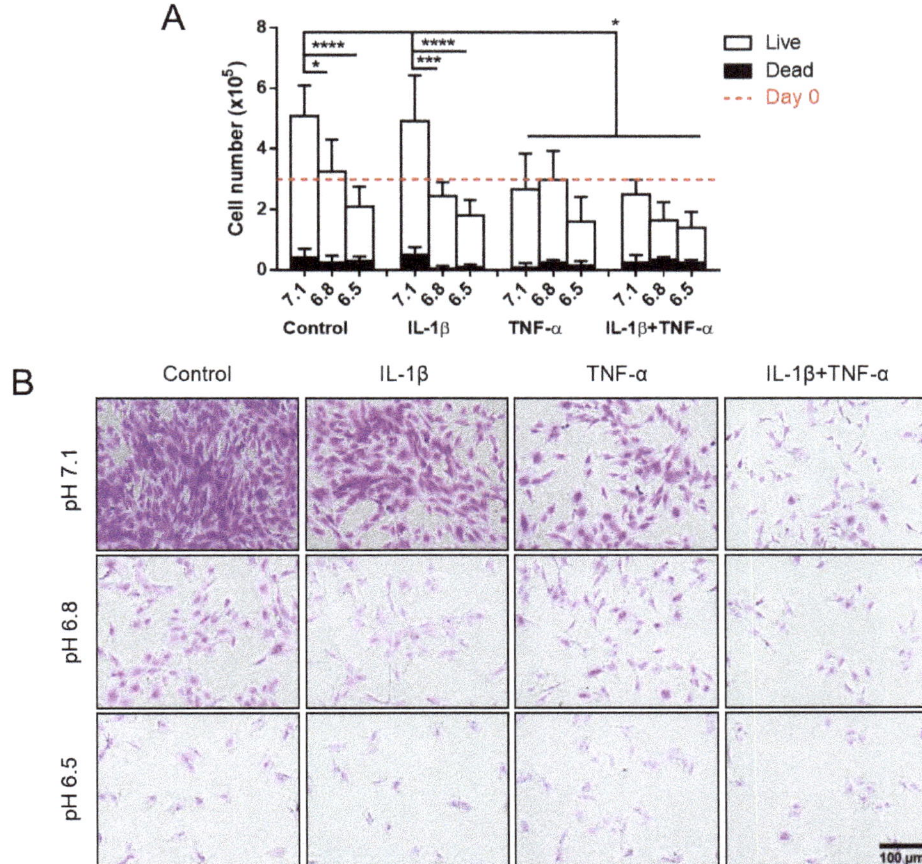

Figure 1. 2D culture effects of pH and inflammatory cytokines on nasal chondrocytes (NCs). (**A**) Viable cell count at day 7 for NCs on 2D tissue culture plastic cultured in pH modified media (pH 7.1, 6.8 and 6.5) either unsupplemented (control) or supplemented with interleukin-1ß (IL1-β), tumour necrosis factor-α (TNF-α) or a combination of both. * ($p < 0.05$), *** ($p < 0.001$) and **** ($p < 0.0001$) indicate significant differences between groups. N = 1 donor, n = 3 samples. (**B**) Crystal violet staining of NCs at day 7. Scale bar is 100 µm.

3.2. Acidic pH and Inflammatory Cytokines Negatively Impact Cell Viability in 3D Hydrogels

When cultured in 3D hydrogels, NCs appeared to be affected by culture conditions in a similar fashion to those observed for 2D culture. Semi-quantitative analysis based on Live/Dead images of hydrogels at day 14 showed a correlation between culture conditions and cell viability with acidic pH, and on a smaller scale the presence of inflammatory cytokines, negatively influencing cell viability (Figure 2A,B). At pH 6.5, supplementation with IL-1β+TNF-α had a more detrimental effect on cell viability than supplementation with single cytokines. However, this result was not confirmed by DNA quantification, with no significant differences observed among sub-groups at pH 6.5 and 6.8 (Figure 2C). Nevertheless, DNA content was found to be lower for groups cultured in acidic pH media, irrespective of the presence of inflammatory cytokines, with the lowest DNA content observed for all pH 6.5 media formulations.

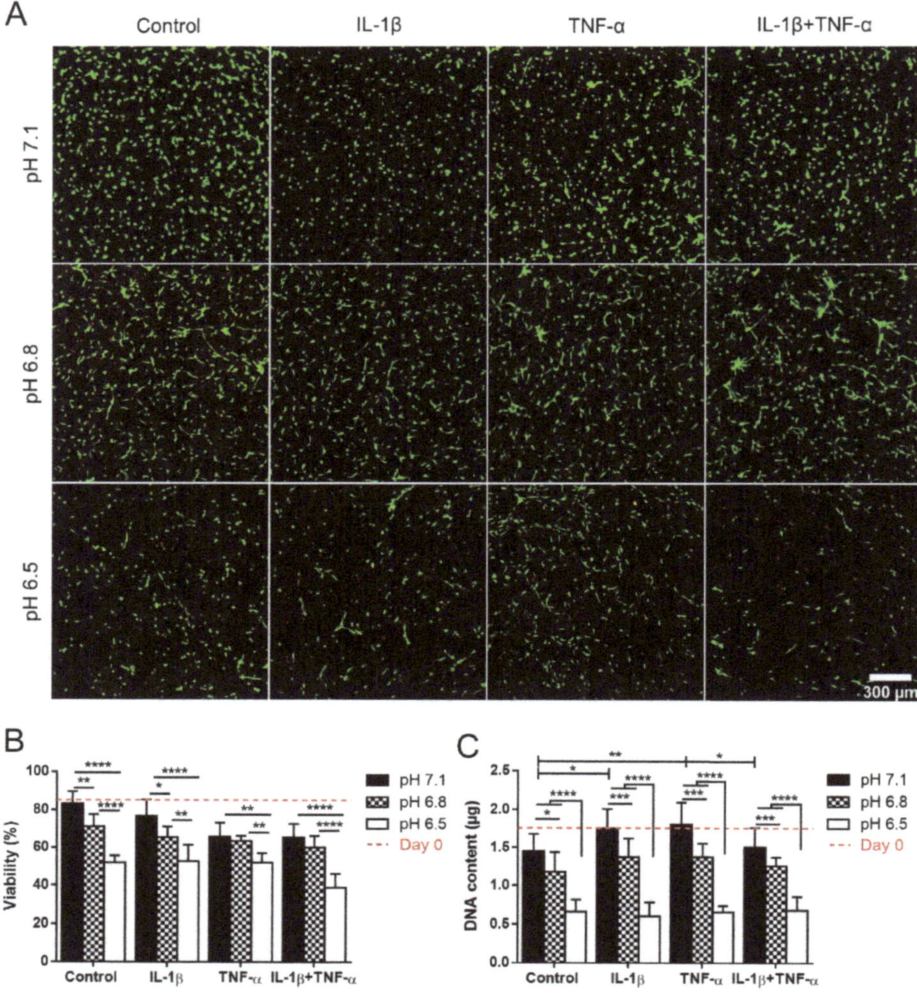

Figure 2. (**A**) Live/dead images of NCs at day 14 cultured in 3D hydrogels in pH modified media (pH 7.1, 6.8 and 6.5) either unsupplemented (control) or supplemented with IL1-β and TNF-α or a combination of both. Scale bar is 300 μm. (**B**) Cell viability (%). (**C**) DNA content (μg) of hydrogels at day 14. * ($p < 0.05$), ** ($p < 0.01$), *** ($p < 0.001$) and **** ($p < 0.0001$) indicate significant differences between groups. N = 3 donors, n = 3 samples.

3.3. Matrix Synthesis in Response to Different pH and Inflammatory Cytokine Conditions

Control ECM hydrogels (no inflammatory cytokines) cultured in media at pH 7.1 exhibited the highest collagen accumulation which was diminished with increasing acidic media formulations. Similar collagen levels were observed for all inflammatory cytokine supplemented groups at all pH levels. At pH 7.1 a noticeable decrease in collagen deposition was observed for IL-1β+TNF-α supplemented formulations compared to control (Figure 3A). The majority of collagen detected at day 14 was found to have been released into the culture media (Figure 3B). Immunohistochemical staining of collagen types I and II did not reveal obvious differences in pericellular collagen deposition across the groups (Figure 3C).

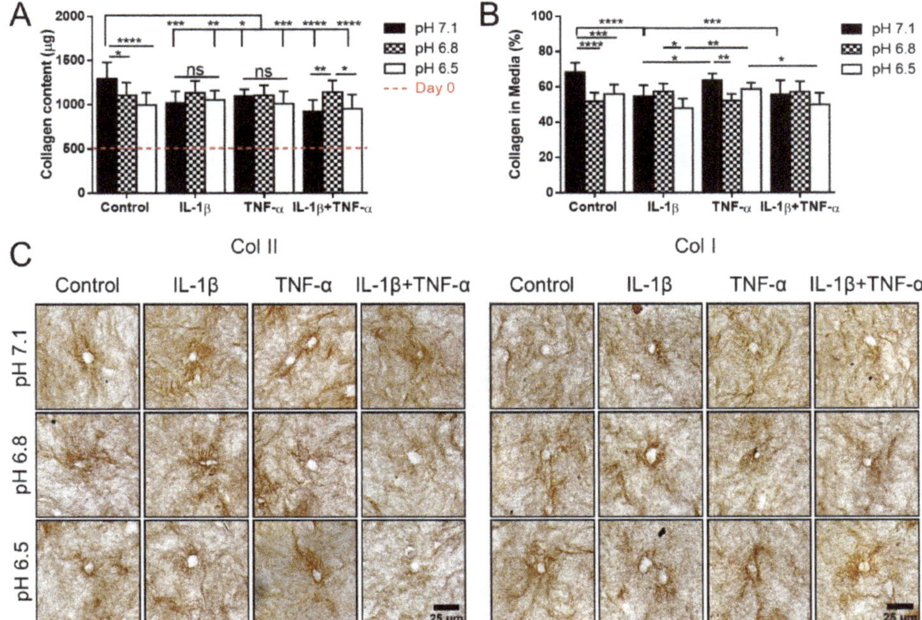

Figure 3. (**A**) Total collagen content (µg) in hydrogels and media at day 14 cultured in pH modified media (pH 7.1, 6.8 and 6.5) either unsupplemented (control) or supplemented with IL1-β and TNF-α or a combination of both. (**B**) Collagen detected in culture media (%). The values are expressed as a percentage of the total amount of collagen measured at day 14. * ($p < 0.05$), ** ($p < 0.01$), *** ($p < 0.001$) and **** ($p < 0.0001$) indicate significant differences between groups, ns (not significant) indicates $p \geq 0.05$. N = 3 donors, n = 3 samples. (**C**) Collagen type II and collagen type I immunohistochemical staining. Scale bar is 25µm.

Acidity was observed to decrease the ability of cells to synthesise sGAGs, while inflammatory cytokine supplementation did not have any noticeable effect (Figure 4A). All pH 7.1 groups exhibited a sGAG to collagen ratio that was significantly higher compared to pH 6.8 and pH 6.5 groups. Interestingly, significant differences in sGAG:collagen ratios were noted among groups cultured at pH 7.1 supplemented with different combinations of inflammatory cytokines. Groups cultured at pH 7.1 supplemented with IL-1β and IL-1β+TNF-α were significantly higher than the pH 7.1 control group (Figure 4B), due to reduced collagen deposition. Histological staining was found to corroborate the sGAG biochemical findings, with more intense staining at pH 7.1 at day 14 (Figure 4C).

3.4. Acidic pH and Inflammatory Cytokines Have a Greater Impact on NP Cell Viability and Proliferation

Motivated by our previous results, the effect of acidic media culture in the presence of physiologically relevant inflammatory cytokine concentrations was subsequently assessed for NCs and NP cells in 3D hydrogel culture. pH media groups were reduced to pH 7.1 and pH 6.5 to simulate healthy and acute degeneration conditions, respectively, with or without the combined supplementation of both IL-1β+TNF-α. Live/dead analysis revealed reduced viability for NP cells compared to NCs at pH 7.1 and pH 6.5 in both control and IL-1β+TNF-α supplemented conditions. Acidity was again observed to be the primary factor responsible for cell death, as shown by the significant decrease in cell viability among cells cultured in acidic media compared to the controls (Figure 5A,B). Quantification of DNA further confirmed the detrimental effect of acidic pH on NCs and NP cells in control and inflamed conditions (Figure 5C).

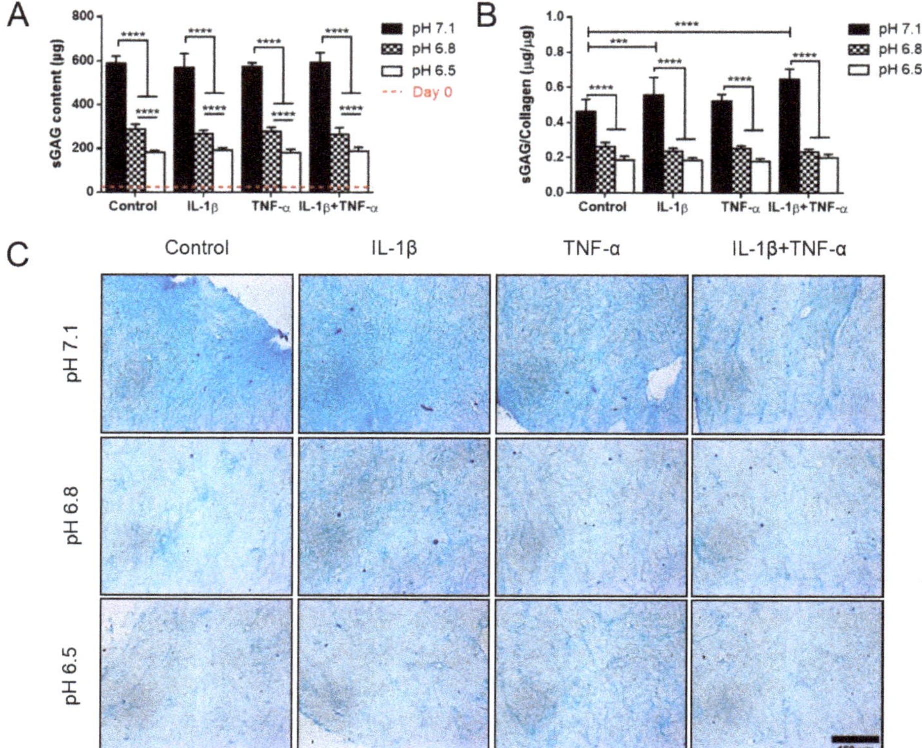

Figure 4. (**A**) Sulphated glycosaminoglycan (sGAG) content (µg) in samples (hydrogels+culture media) at day 14 cultured in pH modified media (pH 7.1, 6.8 and 6.5) either unsupplemented (control) or supplemented with IL1-β and TNF-α or a combination of both. Dashed line represents day 0 sGAG content. (**B**) sGAG:Collagen ratio. Both sGAG and collagen were calculated as respective amounts produced over 14 days by subtracting day 0 content from day 14. *** ($p < 0.001$) and **** ($p < 0.0001$) indicate significant differences between groups. N = 3 donors, n = 3 samples. (**C**) Alcian blue histological evaluation for sGAG accumulation in hydrogels at day 14. Scale bar is 150 µm.

3.5. NCs Can Secrete Higher Amounts of Key Matrix Components Compared to NP Cells in a Degenerated Disc-Like Environment

Acidity was also observed to have a negative impact on sGAG synthesis. Intense localised sGAG deposition was observed for NCs cultured in pH 7.1 media (control and inflamed conditions), with less intense staining observed for NP cells. At pH 6.5, both NCs and NP cells displayed diminished sGAG staining. There were no appreciable differences in staining intensity between control and inflamed conditions at either pH levels. Biochemical results supported these histological findings, demonstrating that NCs cultured in pH 7.1 media were able to synthesise significantly higher amounts of sGAGs compared to all other groups by day 14. There was no significant impact of culturing in IL-1β+TNF-α supplemented media for either cell type (Figure 6A,B).

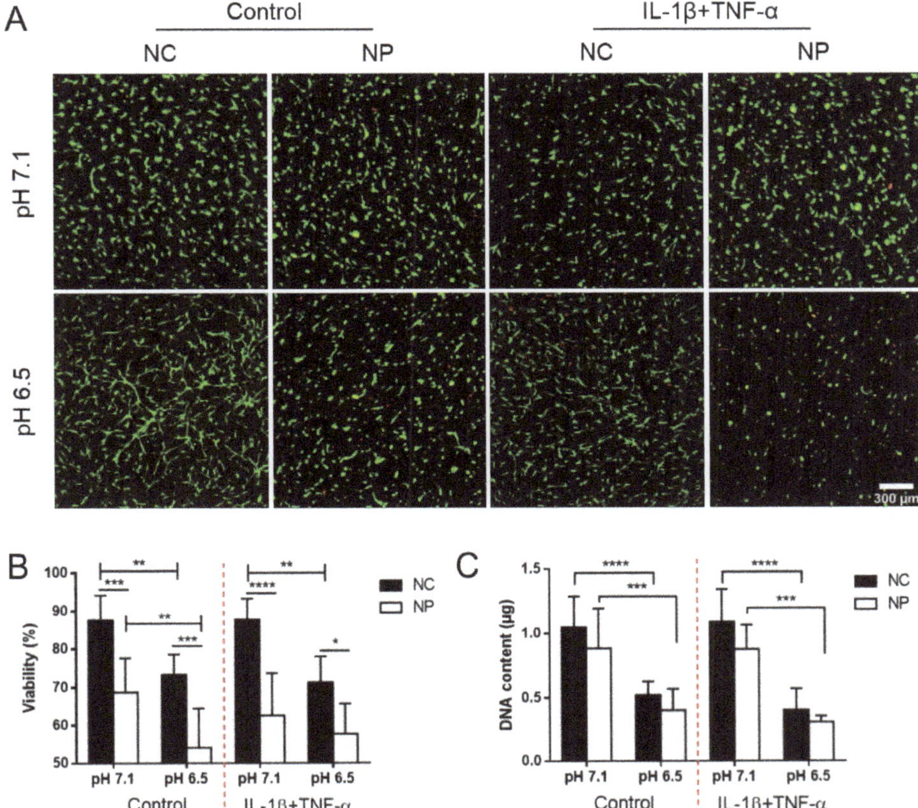

Figure 5. (**A**) Live/dead imaging at day 14 of NCs and NP cells in 3D hydrogel culture at different pH (7.1 and 6.5) and inflammatory cytokine (IL-1β+TNF-α) supplementation conditions. Scale bar is 300 μm. (**B**) Cell viability (%). (**C**) DNA content (μg) of hydrogels at day 14. * ($p < 0.05$), ** ($p < 0.01$), *** ($p < 0.001$) and **** ($p < 0.0001$) indicate significant differences between groups. N = 3 donors, n = 3 samples.

In terms of collagen content, cells cultured in pH 7.1 control media showed enhanced collagen production compared to pH 6.5 control media. However, no differences were detected in collagen synthesis when cells were cultured in IL-1β+TNF-α supplemented media (Figure 7A). Moreover, a similar pattern was found for sGAG:collagen ratios among groups, with NCs exhibiting a higher ratio than NP cells which was higher for groups cultured in pH 7.1 media than pH 6.5. Interestingly, NCs cultured in pH 7.1 media supplemented with IL-1β+TNF-α exhibited a sGAG:collagen ratio significantly higher than in pH 7.1 control media (Figure 7B), due to diminished collagen deposition. Immunohistochemical staining demonstrated higher levels of collagen type II at pH 7.1 which was diminished at pH 6.5. Culturing with IL-1β+TNF-α did not appear to affect the type of collagen being deposited (Figure 7C).

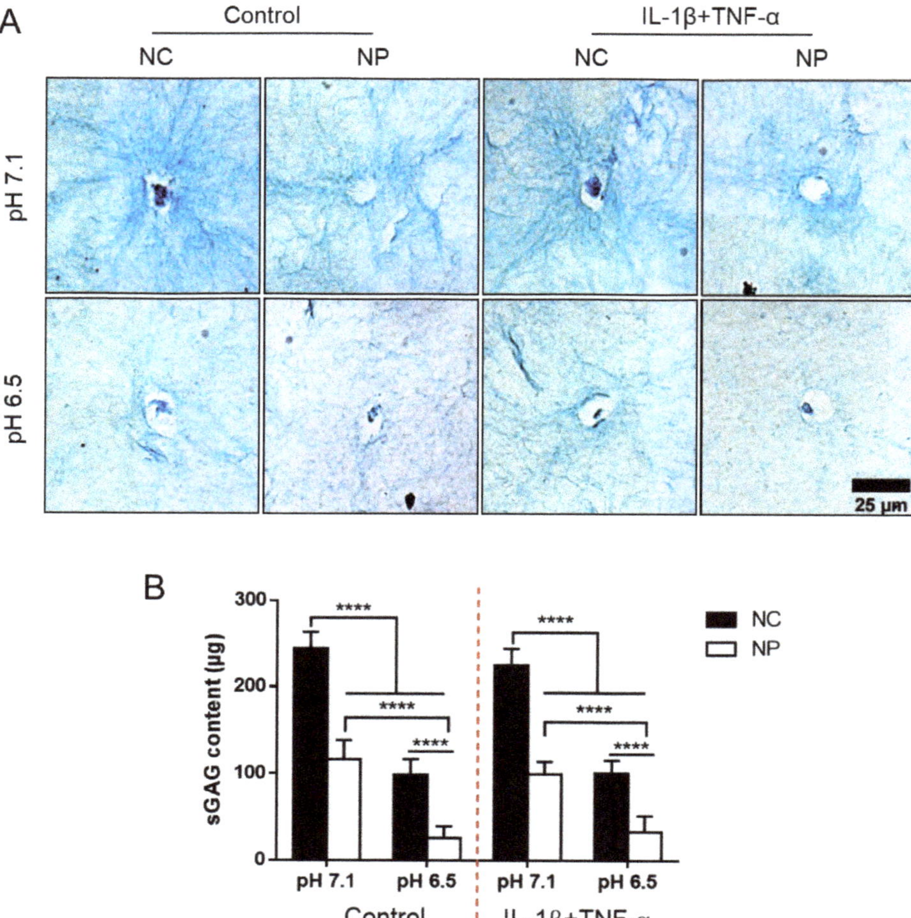

Figure 6. (**A**) Histological evaluation of hydrogels with alcian blue to identify sGAG at day 14 of NCs and NP cells in 3D hydrogel culture at different pH (7.1 and 6.5) and inflammatory cytokine (IL-1β+TNF-α) supplementation conditions. Scale bar is 25 µm. (**B**) sGAG (µg) produced by NCs and NP cells in 3D hydrogels. Values were calculated by subtracting day 0 content from day 14. **** ($p < 0.0001$) indicates significant differences between groups. N = 3 donors, n = 3 samples.

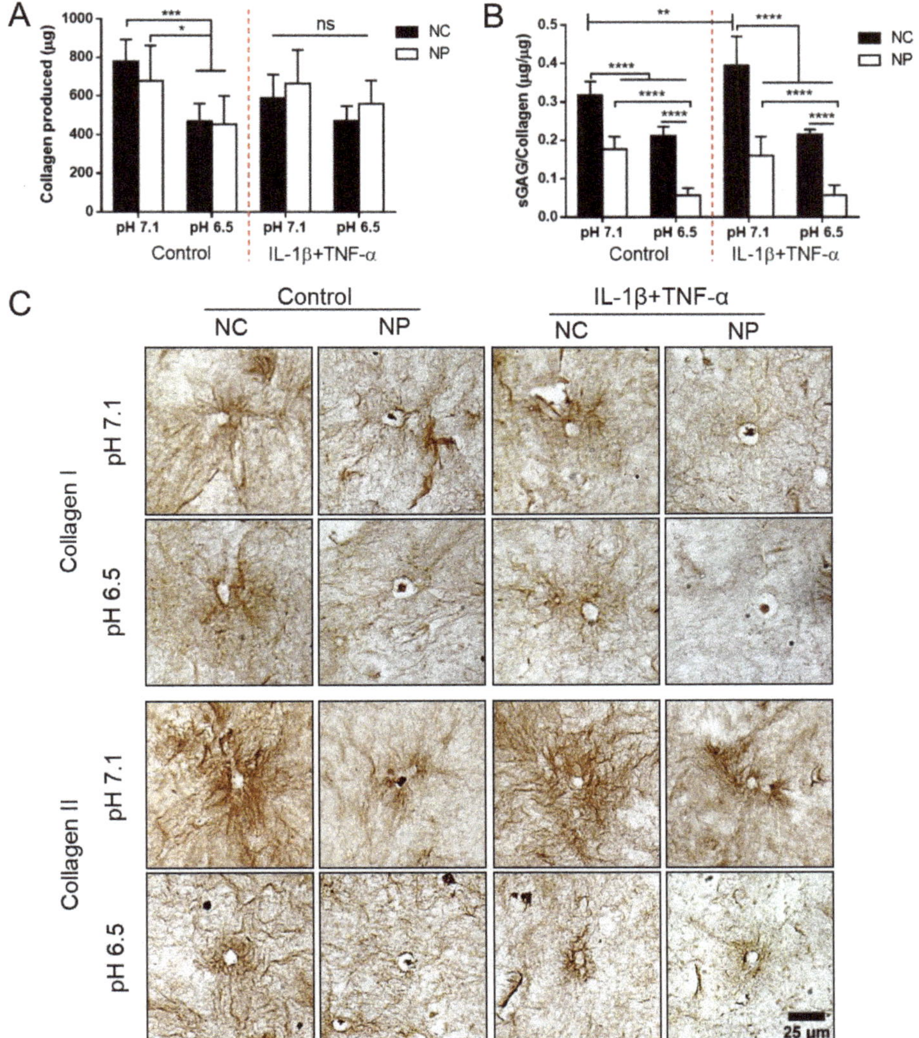

Figure 7. (**A**) Total collagen (µg) produced in hydrogels containing NCs and NP cells in 3D hydrogel culture at different pH (7.1 and 6.5) and inflammatory cytokine (IL-1β+TNF-α) supplementation conditions. (**B**) sGAG:Collagen ratio. Both quantities were calculated as respective amounts produced over 14 days by subtracting day 0 content from day 14. * ($p < 0.05$), ** ($p < 0.01$), *** ($p < 0.001$) and **** ($p < 0.0001$) indicate significant differences between groups, ns (not significant) indicates $p \geq 0.05$. N = 3 donors, n = 3 samples. (**C**) Collagen type I and collagen type II immunohistochemical staining. Scale bar is 25 µm.

4. Discussion

The IVD is a large avascular organ relying mainly on diffusion of nutrients through the CEPs for its energy supply. During disc degeneration calcification of the CEPs hinders the availability of nutrients to cells residing in the centre of the disc and the removal of metabolic by-products of glycolysis, thereby causing acidification of the cellular microenvironment and a shift to an accelerated catabolic state [14,15]. Moreover, the degenerated disc consists of a wide panel of pro-inflammatory cytokines [22,29,43], further contributing to the creation of an extremely challenging biochemical microenvironment. For successful translation of cell-based therapies for disc regeneration, it is imperative that cells remain viable post injection, and function effectively in a challenging microenvironment [8,33,34].

In the first part of this study, we investigated the impact of the pro-inflammatory cytokines IL-1β and TNF-α, in combination with low pH media in 2D and 3D hydrogel culture with NCs. It was found that acidic pH affected cell viability and number to a much greater extent in comparison to IL-1β and TNF-α supplementation. The detrimental effect of acidic pH was also observed in relation to a significant reduction in matrix synthesis in 3D culture, whereas IL-1β and TNF-α supplementation, either alone or in combination, appeared to have no influence on sGAG synthesis and only a marginal influence on collagen production when NCs were cultured at pH 7.1.

Surprisingly, we found that cytokine supplementation had only a marginal influence on cell viability and matrix synthesis. Cellular response to culture in inflammatory conditions has been extensively studied in the last two decades. However, there is still no consensus on the specific molecular mechanisms that are triggered or enhanced by the presence of pro-inflammatory cytokines such as IL-1β and TNF-α, possibly due to the variability in culture conditions and consequently in the results obtained. Some studies have found strong evidence of IL-1β being the key regulator of matrix degradation processes in NP cells [28,44], and identified the role of TNF-α more likely to be associated with nerve root irritation rather than influencing matrix catabolism [23]. In contrast, a number of other studies have either not found major differences between the effects of IL-1β and TNF-α supplementation on NP cells [27,45], or even reconsidered the central role that was attributed to IL-1β in the process of matrix catabolism [19,43]. It is possible to find analogous discrepancies in similar studies on articular chondrocytes (ACs); whereby some authors have judged the contributions of both IL-1β and TNF-α as equal [46], others have found evidence of TNF-α having a predominant influence on cell death [47] and decreased matrix synthesis [48]. Nevertheless, in contrast with our results, the common denominator of all these findings is that the presence of pro-inflammatory cytokines, either alone or in combination, deeply affects cell fate and gene expression. It must be noted, however, that in all these previous studies cells have been exposed to concentrations of IL-1β and TNF-α of the order of ng/mL, which is supraphysiological even for pathological conditions such as the more acute stages of degeneration of the IVD. In a recent study from Altun et al., biopsy specimens from surgically excised discs from patients diagnosed with acute disc disease were shown to contain amounts of IL-1β and TNF-α in the range of 113–135 pg/mL and 9–26 pg/mL, respectively [29], which are more than three orders of magnitude lower than the concentrations used in previous experiments. Our results suggest that when cell cultures are supplemented with IL-1β and TNF-α at physiologically relevant concentrations, cellular responses are not as marked as described in previous experiments and that these cytokines affect cell viability and matrix synthesis less dramatically compared to environmental pH.

The effects of low pH on cell viability have been explored in the last two decades on different cell sources such as bone marrow stem cells (BMSCs) [49], adipose-derived mesenchymal stem cells (ADMSCs) [50], ACs [51,52] and NP cells [53]. Results from these studies are in line with our findings, demonstrating a correlation between increasing environmental acidity and decreased cell viability. One possible explanation for this could be that culture in acidic media induces the activation of acid-sensing ion channels (ASICs), which are extracellular pH sensors whose activation has been found to be responsible for acidosis-mediated apoptosis [54]. The expression of this family of receptors has been observed in several cell types, including disc cells and ACs [51,54–56]. Although the expression of ASICs has not yet been examined for chondrocytes extracted specifically from the nasal septum, it is reasonable to assume that their response to acidic environments would be driven by the same biochemical mechanisms that drives the response of chondrocytes derived from articular cartilage. Activation of ASICs has been shown to be enhanced under inflammatory conditions, in particular in the presence of IL-1β, resulting in increased cell death [57]. However, such a correlation was not observed in our results, which showed no significant differences in cell viability following IL-1β supplementation in 2D or 3D cultures.

Acidic pH also had a significant impact on the ability of NCs to synthesise and deposit de novo matrix. We observed a reduction in the synthesis of sGAGs compared to controls (pH 7.1) in NCs cultured in media at pH 6.8 and 6.5, which is not simply due to compromised cell viability. It has previously been reported that cell metabolism can be severely inhibited by low pH [58,59]. This is due to phosphofructokinases, the rate-limiting enzymes on glycolysis, being sensitive to pH and in particular inhibited by acidosis [60]. As a consequence, cells have less energy available and processes tightly linked to energy consumption such as sGAG synthesis are inhibited or impeded [61]. Interestingly, although we did observe a higher amount of collagen being produced at pH 7.1 without inflammatory factors, the detrimental impact of acidic conditions was less pronounced in comparison to sGAG synthesis, suggesting that the molecular mechanisms regulating collagen synthesis are less dependent on energy availability. A similar finding was reported in a study from Nishida et al., where a decrease in energy production in hypoxic conditions promoted hypertrophy in ACs with a decrease in aggrecan and an increase in collagen expression [62].

A second objective of this study was to directly compare the ability of NCs and NP cells to synthesise NP-like matrix in culture conditions mimicking the degenerated disc niche (low glucose, low oxygen, low pH in the presence of pro-inflammatory cytokines), with the aim of determining an appropriate cell source for disc repair strategies. We found that NCs, although still negatively affected by low pH, maintained higher cell viability and synthesised appropriate matrix components such as sGAGs and collagens in a more appropriate ratio than NP cells. In comparison, NP cells exhibited diminished cell viability at pH 7.1, representative of a healthy disc, which also resulted in a reduction in the accumulation of sGAGs when compared to NCs. This correlates with results from Razaq et al., where the expression profile of bovine AC and NP cells cultured in acidic conditions was compared, demonstrating a significantly lower expression of matrix related genes with increasing acidity for NP cells [53]. Importantly, the sGAG:collagen ratio was considerably higher in NCs compared to NP cells for all culture conditions. A high sGAG:collagen ratio may be considered an appropriate parameter for the identification of NP-like tissue type, as it represents a tissue rich in sGAGs, replicating the composition typical of healthy NP [63].

In conclusion, the results of the present study demonstrate that among the plethora of characteristics that render the degenerated NP an extremely hostile niche, environmental pH was found to have the greatest impact on cell viability and matrix production. While NP cells appeared to be unaffected by the presence of inflammatory cytokines in the physiological range, low pH had a greater impact on viability and matrix synthesis, suggesting they may be a less effective cell type compared to NCs for cell-based NP regeneration. Although it is clear that environmental acidity is an important factor to consider in the context of disc regeneration, to the best of the authors knowledge, there are no clinically feasible approaches to neutralising or augmenting the intradiscal pH. A possible approach may involve injecting pH buffering biomaterials into the disc space. This may have some short-term benefit and create a permissible microenvironment to facilitate regeneration by transplanted cells, but the longevity of such an approach would need to be established. In parallel, techniques to non-invasively measure intradiscal pH in vivo would be required in order to tailor or adjust the treatment according to the severity of the degenerated state. Recent work has proposed a promising clinical MRI approach to detect pH changes associated with IVD degeneration in a swine model [64]. Overall, the presence of the key inflammatory cytokines IL-1β and TNF-α, that have previously been identified in human discs did not have any appreciable impact when supplemented in physiologically relevant concentrations. Therefore, developing imaging modalities to identify and characterise intradiscal pH in vivo in combination with pH buffering biomaterials may provide for an effective strategy to appropriately select and treat patients with cell-based therapies.

Author Contributions: Both the authors (C.B. and C.T.B.) have contributed equally to this manuscript and have read and agreed to the published version of the manuscript. All authors have read and agreed to the published version of the manuscript.

Funding: This work was supported by Science Foundation Ireland Career Development Award (15/CDA/3476).

Conflicts of Interest: The authors have no conflict of interest to declare.

References

1. Vos, T.; Flaxman, A.D.; Naghavi, M.; Lozano, R.; Michaud, C.; Ezzati, M.; Shibuya, K.; Salomon, J.A.; Abdalla, S.; Aboyans, V.; et al. Years lived with disability (YLDs) for 1160 sequelae of 289 diseases and injuries 1990-2010: A systematic analysis for the Global Burden of Disease Study 2010. *Lancet* **2012**, *380*, 2163–2196. [CrossRef]
2. Magnusson, M.L.; Aleksiev, A.; Wilder, D.G.; Pope, M.H.; Spratt, K.; Lee, S.H.; Goel, V.K.; Weinstein, J.N. European Spine Society–the AcroMed Prize for Spinal Research 1995. Unexpected load and asymmetric posture as etiologic factors in low back pain. *Eur. Spine J.* **1996**, *5*, 23–35. [CrossRef] [PubMed]
3. Patel, A.A.; Spiker, W.R.; Daubs, M.; Brodke, D.; Cannon-Albright, L.A. Evidence for an inherited predisposition to lumbar disc disease. *J. Bone Jt. Surg. Am.* **2011**, *93*, 225–229. [CrossRef] [PubMed]
4. Lefevre-Colau, M.M.; Fayad, F.; Rannou, F.; Fermanian, J.; Coriat, F.; Mace, Y.; Revel, M.; Poiraudeau, S. Frequency and interrelations of risk factors for chronic low back pain in a primary care setting. *PLoS ONE* **2009**, *4*, e4874. [CrossRef]
5. Vergroesen, P.P.; Kingma, I.; Emanuel, K.S.; Hoogendoorn, R.J.; Welting, T.J.; van Royen, B.J.; van Dieen, J.H.; Smit, T.H. Mechanics and biology in intervertebral disc degeneration: A vicious circle. *Osteoarthr. Cartil.* **2015**, *23*, 1057–1070. [CrossRef]
6. Luoma, K.; Riihimaki, H.; Luukkonen, R.; Raininko, R.; Viikari-Juntura, E.; Lamminen, A. Low back pain in relation to lumbar disc degeneration. *Spine* **2000**, *25*, 487–492. [CrossRef]
7. Urban, J.P.; Smith, S.; Fairbank, J.C. Nutrition of the intervertebral disc. *Spine* **2004**, *29*, 2700–2709. [CrossRef] [PubMed]
8. Buckley, C.T.; Hoyland, J.A.; Fujii, K.; Pandit, A.; Iatridis, J.C.; Grad, S. Critical aspects and challenges for intervertebral disc repair and regeneration-Harnessing advances in tissue engineering. *JOR Spine* **2018**, *1*, e1029. [CrossRef]
9. Grant, M.P.; Epure, L.M.; Bokhari, R.; Roughley, P.; Antoniou, J.; Mwale, F. Human cartilaginous endplate degeneration is induced by calcium and the extracellular calcium-sensing receptor in the intervertebral disc. *Eur. Cell Mater.* **2016**, *32*, 137–151. [CrossRef]
10. Tomaszewski, K.A.; Adamek, D.; Konopka, T.; Tomaszewska, R.; Walocha, J.A. Endplate calcification and cervical intervertebral disc degeneration: The role of endplate marrow contact channel occlusion. *Folia Morphol.* **2015**, *74*, 84–92. [CrossRef]
11. Bartels, E.M.; Fairbank, J.C.; Winlove, C.P.; Urban, J.P. Oxygen and lactate concentrations measured in vivo in the intervertebral discs of patients with scoliosis and back pain. *Spine* **1998**, *23*, 1–7; discussion 8. [CrossRef] [PubMed]
12. Nachemson, A. Intradiscal measurements of pH in patients with lumbar rhizopathies. *Acta Orthop. Scand.* **1969**, *40*, 23–42. [CrossRef] [PubMed]
13. Sivan, S.S.; Hayes, A.J.; Wachtel, E.; Caterson, B.; Merkher, Y.; Maroudas, A.; Brown, S.; Roberts, S. Biochemical composition and turnover of the extracellular matrix of the normal and degenerate intervertebral disc. *Eur. Spine J.* **2014**, *23* (Suppl. 3), S344–S353. [CrossRef] [PubMed]
14. Urban, J.P. The role of the physicochemical environment in determining disc cell behaviour. *Biochem. Soc. Trans.* **2002**, *30*, 858–864. [CrossRef]
15. Hadjipavlou, A.G.; Tzermiadianos, M.N.; Bogduk, N.; Zindrick, M.R. The pathophysiology of disc degeneration: A critical review. *J. Bone Jt. Surg. Br.* **2008**, *90*, 1261–1270. [CrossRef]
16. Green, J.D.; Tollemar, V.; Dougherty, M.; Yan, Z.; Yin, L.; Ye, J.; Collier, Z.; Mohammed, M.K.; Haydon, R.C.; Luu, H.H.; et al. Multifaceted signaling regulators of chondrogenesis: Implications in cartilage regeneration and tissue engineering. *Genes Dis.* **2015**, *2*, 307–327. [CrossRef]
17. Molinos, M.; Almeida, C.R.; Caldeira, J.; Cunha, C.; Goncalves, R.M.; Barbosa, M.A. Inflammation in intervertebral disc degeneration and regeneration. *J. R. Soc. Interface* **2015**, *12*, 20141191. [CrossRef]
18. Shamji, M.F.; Setton, L.A.; Jarvis, W.; So, S.; Chen, J.; Jing, L.; Bullock, R.; Isaacs, R.E.; Brown, C.; Richardson, W.J. Proinflammatory cytokine expression profile in degenerated and herniated human intervertebral disc tissues. *Arthritis Rheum.* **2010**, *62*, 1974–1982. [CrossRef]

19. Purmessur, D.; Walter, B.A.; Roughley, P.J.; Laudier, D.M.; Hecht, A.C.; Iatridis, J. A role for TNFalpha in intervertebral disc degeneration: A non-recoverable catabolic shift. *Biochem. Biophys. Res. Commun.* **2013**, *433*, 151–156. [CrossRef]
20. Shen, C.; Yan, J.; Jiang, L.S.; Dai, L.Y. Autophagy in rat annulus fibrosus cells: Evidence and possible implications. *Arthritis Res. Ther.* **2011**, *13*, R132. [CrossRef] [PubMed]
21. Johnson, Z.I.; Schoepflin, Z.R.; Choi, H.; Shapiro, I.M.; Risbud, M.V. Disc in flames: Roles of TNF-alpha and IL-1beta in intervertebral disc degeneration. *Eur. Cell Mater.* **2015**, *30*, 104–116; discussion 116-107. [CrossRef] [PubMed]
22. Risbud, M.V.; Shapiro, I.M. Role of cytokines in intervertebral disc degeneration: Pain and disc content. *Nat. Rev. Rheumatol.* **2014**, *10*, 44–56. [CrossRef] [PubMed]
23. Hoyland, J.A.; Le Maitre, C.; Freemont, A.J. Investigation of the role of IL-1 and TNF in matrix degradation in the intervertebral disc. *Rheumatology* **2008**, *47*, 809–814. [CrossRef] [PubMed]
24. Yang, W.; Yu, X.H.; Wang, C.; He, W.S.; Zhang, S.J.; Yan, Y.G.; Zhang, J.; Xiang, Y.X.; Wang, W.J. Interleukin-1beta in intervertebral disk degeneration. *Clin. Chim. Acta* **2015**, *450*, 262–272. [CrossRef] [PubMed]
25. Hayashi, S.; Taira, A.; Inoue, G.; Koshi, T.; Ito, T.; Yamashita, M.; Yamauchi, K.; Suzuki, M.; Takahashi, K.; Ohtori, S. TNF-alpha in nucleus pulposus induces sensory nerve growth: A study of the mechanism of discogenic low back pain using TNF-alpha-deficient mice. *Spine* **2008**, *33*, 1542–1546. [CrossRef] [PubMed]
26. Le Maitre, C.L.; Freemont, A.J.; Hoyland, J.A. The role of interleukin-1 in the pathogenesis of human intervertebral disc degeneration. *Arthritis Res. Ther.* **2005**, *7*, R732–R745. [CrossRef]
27. Wang, J.; Markova, D.; Anderson, D.G.; Zheng, Z.; Shapiro, I.M.; Risbud, M.V. TNF-alpha and IL-1beta promote a disintegrin-like and metalloprotease with thrombospondin type I motif-5-mediated aggrecan degradation through syndecan-4 in intervertebral disc. *J. Biol. Chem.* **2011**, *286*, 39738–39749. [CrossRef]
28. Le Maitre, C.L.; Hoyland, J.A.; Freemont, A.J. Interleukin-1 receptor antagonist delivered directly and by gene therapy inhibits matrix degradation in the intact degenerate human intervertebral disc: An in situ zymographic and gene therapy study. *Arthritis Res. Ther.* **2007**, *9*, R83. [CrossRef]
29. Altun, I. Cytokine profile in degenerated painful intervertebral disc: Variability with respect to duration of symptoms and type of disease. *Spine J.* **2016**, *16*, 857–861. [CrossRef]
30. Mathy-Hartert, M.; Deby-Dupont, G.P.; Reginster, J.Y.; Ayache, N.; Pujol, J.P.; Henrotin, Y.E. Regulation by reactive oxygen species of interleukin-1beta, nitric oxide and prostaglandin E(2) production by human chondrocytes. *Osteoarthr. Cartil.* **2002**, *10*, 547–555. [CrossRef]
31. Rannou, F.; Corvol, M.T.; Hudry, C.; Anract, P.; Dumontier, M.F.; Tsagris, L.; Revel, M.; Poiraudeau, S. Sensitivity of anulus fibrosus cells to interleukin 1 beta. Comparison with articular chondrocytes. *Spine* **2000**, *25*, 17–23. [CrossRef] [PubMed]
32. Wang, J.; Tian, Y.; Phillips, K.L.; Chiverton, N.; Haddock, G.; Bunning, R.A.; Cross, A.K.; Shapiro, I.M.; Le Maitre, C.L.; Risbud, M.V. Tumor necrosis factor alpha- and interleukin-1beta-dependent induction of CCL3 expression by nucleus pulposus cells promotes macrophage migration through CCR1. *Arthritis Rheum.* **2013**, *65*, 832–842. [CrossRef] [PubMed]
33. Vedicherla, S.; Buckley, C.T. Cell-based therapies for intervertebral disc and cartilage regeneration- Current concepts, parallels, and perspectives. *J. Orthop. Res.* **2017**, *35*, 8–22. [CrossRef] [PubMed]
34. Smith, L.J.; Silverman, L.; Sakai, D.; Le Maitre, C.L.; Mauck, R.L.; Malhotra, N.R.; Lotz, J.C.; Buckley, C.T. Advancing cell therapies for intervertebral disc regeneration from the lab to the clinic: Recommendations of the ORS spine section. *JOR Spine* **2018**, *1*, e1036. [CrossRef] [PubMed]
35. Mumme, M.; Barbero, A.; Miot, S.; Wixmerten, A.; Feliciano, S.; Wolf, F.; Asnaghi, A.M.; Baumhoer, D.; Bieri, O.; Kretzschmar, M.; et al. Nasal chondrocyte-based engineered autologous cartilage tissue for repair of articular cartilage defects: An observational first-in-human trial. *Lancet* **2016**, *388*, 1985–1994. [CrossRef]
36. Chen, W.; Li, C.; Peng, M.; Xie, B.; Zhang, L.; Tang, X. Autologous nasal chondrocytes delivered by injectable hydrogel for in vivo articular cartilage regeneration. *Cell Tissue Bank.* **2018**, *19*, 35–46. [CrossRef]
37. Kafienah, W.; Jakob, M.; Demarteau, O.; Frazer, A.; Barker, M.D.; Martin, I.; Hollander, A.P. Three-dimensional tissue engineering of hyaline cartilage: Comparison of adult nasal and articular chondrocytes. *Tissue Eng.* **2002**, *8*, 817–826. [CrossRef]
38. Vedicherla, S.; Buckley, C.T. In vitro extracellular matrix accumulation of nasal and articular chondrocytes for intervertebral disc repair. *Tissue Cell* **2017**, *49*, 503–513. [CrossRef]

39. Vedicherla, S.; Buckley, C.T. Rapid Chondrocyte Isolation for Tissue Engineering Applications: The Effect of Enzyme Concentration and Temporal Exposure on the Matrix Forming Capacity of Nasal Derived Chondrocytes. *Biomed. Res. Int.* **2017**, *2017*, 2395138. [CrossRef]
40. Borrelli, C.; Buckley, C.T. Injectable Disc-Derived ECM Hydrogel Functionalised with Chondroitin Sulfate for Intervertebral Disc Regeneration. *Acta Biomater.* **2020**, *117*, 142–155. [CrossRef]
41. Kafienah, W.; Sims, T.J. Biochemical methods for the analysis of tissue-engineered cartilage. *Methods Mol. Biol.* **2004**, *238*, 217–230. [PubMed]
42. Ignat'eva, N.Y.; Danilov, N.A.; Averkiev, S.V.; Obrezkova, M.V.; Lunin, V.V.; Sobol, E.N. Determination of hydroxyproline in tissues and the evaluation of the collagen content of the tissues. *J. Anal. Chem.* **2017**, *62*, 51–57. [CrossRef]
43. Phillips, K.L.; Cullen, K.; Chiverton, N.; Michael, A.L.; Cole, A.A.; Breakwell, L.M.; Haddock, G.; Bunning, R.A.; Cross, A.K.; Le Maitre, C.L. Potential roles of cytokines and chemokines in human intervertebral disc degeneration: Interleukin-1 is a master regulator of catabolic processes. *Osteoarthr. Cartil.* **2015**, *23*, 1165–1177. [CrossRef] [PubMed]
44. Srivastava, A.; Isa, I.L.; Rooney, P.; Pandit, A. Bioengineered three-dimensional diseased intervertebral disc model revealed inflammatory crosstalk. *Biomaterials* **2017**, *123*, 127–141. [CrossRef]
45. Ponnappan, R.K.; Markova, D.Z.; Antonio, P.J.; Murray, H.B.; Vaccaro, A.R.; Shapiro, I.M.; Anderson, D.G.; Albert, T.J.; Risbud, M.V. An organ culture system to model early degenerative changes of the intervertebral disc. *Arthritis Res. Ther.* **2011**, *13*, R171. [CrossRef]
46. Xiang, Y.; Masuko-Hongo, K.; Sekine, T.; Nakamura, H.; Yudoh, K.; Nishioka, K.; Kato, T. Expression of proteinase-activated receptors (PAR)-2 in articular chondrocytes is modulated by IL-1beta, TNF-alpha and TGF-beta. *Osteoarthr. Cartil.* **2006**, *14*, 1163–1173. [CrossRef]
47. Carames, B.; Lopez-Armada, M.J.; Cillero-Pastor, B.; Lires-Dean, M.; Vaamonde, C.; Galdo, F.; Blanco, F.J. Differential effects of tumor necrosis factor-alpha and interleukin-1beta on cell death in human articular chondrocytes. *Osteoarthr. Cartil.* **2008**, *16*, 715–722. [CrossRef]
48. Kuroki, K.; Stoker, A.M.; Cook, J.L. Effects of proinflammatory cytokines on canine articular chondrocytes in a three-dimensional culture. *Am. J. Vet. Res.* **2005**, *66*, 1187–1196. [CrossRef]
49. Naqvi, S.M.; Buckley, C.T. Bone Marrow Stem Cells in Response to Intervertebral Disc-Like Matrix Acidity and Oxygen Concentration: Implications for Cell-based Regenerative Therapy. *Spine* **2016**, *41*, 743–750. [CrossRef]
50. Han, B.; Wang, H.C.; Li, H.; Tao, Y.Q.; Liang, C.Z.; Li, F.C.; Chen, G.; Chen, Q.X. Nucleus pulposus mesenchymal stem cells in acidic conditions mimicking degenerative intervertebral discs give better performance than adipose tissue-derived mesenchymal stem cells. *Cells Tissues Organs* **2014**, *199*, 342–352. [CrossRef]
51. Hu, W.; Chen, F.H.; Yuan, F.L.; Zhang, T.Y.; Wu, F.R.; Rong, C.; Jiang, S.; Tang, J.; Zhang, C.C.; Lin, M.Y. Blockade of acid-sensing ion channels protects articular chondrocytes from acid-induced apoptotic injury. *Inflamm. Res.* **2012**, *61*, 327–335. [CrossRef] [PubMed]
52. Wilkins, R.J.; Hall, A.C. Control of matrix synthesis in isolated bovine chondrocytes by extracellular and intracellular pH. *J. Cell Physiol.* **1995**, *164*, 474–481. [CrossRef] [PubMed]
53. Razaq, S.; Wilkins, R.J.; Urban, J.P. The effect of extracellular pH on matrix turnover by cells of the bovine nucleus pulposus. *Eur. Spine J.* **2003**, *12*, 341–349. [CrossRef] [PubMed]
54. Zhou, R.P.; Wu, X.S.; Wang, Z.S.; Xie, Y.Y.; Ge, J.F.; Chen, F.H. Novel Insights into Acid-Sensing Ion Channels: Implications for Degenerative Diseases. *Aging Dis.* **2016**, *7*, 491–501. [CrossRef]
55. Yuan, F.L.; Chen, F.H.; Lu, W.G.; Li, X.; Wu, F.R.; Li, J.P.; Li, C.W.; Wang, Y.; Zhang, T.Y.; Hu, W. Acid-sensing ion channel 1a mediates acid-induced increases in intracellular calcium in rat articular chondrocytes. *Mol. Cell Biochem.* **2010**, *340*, 153–159. [CrossRef]
56. Gilbert, H.T.J.; Hodson, N.; Baird, P.; Richardson, S.M.; Hoyland, J.A. Acidic pH promotes intervertebral disc degeneration: Acid-sensing ion channel -3 as a potential therapeutic target. *Sci. Rep.* **2016**, *6*, 37360. [CrossRef]
57. Gong, W.; Kolker, S.J.; Usachev, Y.; Walder, R.Y.; Boyle, D.L.; Firestein, G.S.; Sluka, K.A. Acid-sensing ion channel 3 decreases phosphorylation of extracellular signal-regulated kinases and induces synoviocyte cell death by increasing intracellular calcium. *Arthritis Res. Ther.* **2014**, *16*, R121. [CrossRef]

58. Das, R.H.; van Osch, G.J.; Kreukniet, M.; Oostra, J.; Weinans, H.; Jahr, H. Effects of individual control of pH and hypoxia in chondrocyte culture. *J. Orthop. Res.* **2010**, *28*, 537–545. [CrossRef]
59. Wu, M.H.; Urban, J.P.; Cui, Z.F.; Cui, Z.; Xu, X. Effect of extracellular ph on matrix synthesis by chondrocytes in 3D agarose gel. *Biotechnol. Prog.* **2007**, *23*, 430–434. [CrossRef]
60. Erecinska, M.; Deas, J.; Silver, I.A. The effect of pH on glycolysis and phosphofructokinase activity in cultured cells and synaptosomes. *J. Neurochem.* **1995**, *65*, 2765–2772. [CrossRef]
61. Lee, R.B.; Urban, J.P. Evidence for a negative Pasteur effect in articular cartilage. *Biochem. J.* **1997**, *321 Pt 1*, 95–102. [CrossRef] [PubMed]
62. Nishida, T.; Kubota, S.; Aoyama, E.; Takigawa, M. Impaired glycolytic metabolism causes chondrocyte hypertrophy-like changes via promotion of phospho-Smad1/5/8 translocation into nucleus. *Osteoarthr. Cartil.* **2013**, *21*, 700–709. [CrossRef] [PubMed]
63. Mwale, F.; Roughley, P.; Antoniou, J. Distinction between the extracellular matrix of the nucleus pulposus and hyaline cartilage: A requisite for tissue engineering of intervertebral disc. *Eur. Cell Mater.* **2004**, *8*, 58–63; discussion 63-54. [CrossRef] [PubMed]
64. Bez, M.; Zhou, Z.; Sheyn, D.; Tawackoli, W.; Giaconi, J.C.; Shapiro, G.; Ben David, S.; Gazit, Z.; Pelled, G.; Li, D.; et al. Molecular pain markers correlate with pH-sensitive MRI signal in a pig model of disc degeneration. *Sci. Rep.* **2018**, *8*, 17363. [CrossRef]

Article

Angiotensin II Type 1 Receptor Antagonist Losartan Inhibits TNF-α-Induced Inflammation and Degeneration Processes in Human Nucleus Pulposus Cells

Babak Saravi [1,2,†], Zhen Li [1,†], Judith Pfannkuche [1,2,†], Laura Wystrach [2], Sonja Häckel [3], Christoph E. Albers [3], Sibylle Grad [1], Mauro Alini [1], Robert Geoffrey Richards [1], Corinna Lang [4,5], Norbert Südkamp [2], Hagen Schmal [2] and Gernot Lang [2,*]

1 AO Research Institute Davos, Clavadelerstrasse 8, 7270 Davos, Switzerland; babak.saravi@jupiter.uni-freiburg.de (B.S.); zhen.li@aofoundation.org (Z.L.); jjpfannkuche@gmail.com (J.P.); sibylle.grad@aofoundation.org (S.G.); mauro.alini@aofoundation.org (M.A.); geoff.richards@aofoundation.org (R.G.R.)

2 Department of Orthopedics and Trauma Surgery, Medical Centre—Albert-Ludwigs-University of Freiburg, Faculty of Medicine, Albert-Ludwigs-University of Freiburg, Hugstetterstrasse 55, 79106 Freiburg, Germany; laura.wystrach@uniklinik-freiburg.de (L.W.); norbert.suedkamp@uniklinik-freiburg.de (N.S.); hagen.schmal@uniklinik-freiburg.de (H.S.)

3 Department of Orthopaedic Surgery and Traumatology, Inselspital, Bern University Hospital, University of Bern, Freiburgstrasse 18, 3010 Bern, Switzerland; Sonja.haeckel@insel.ch (S.H.); Christoph.Albers@insel.ch (C.E.A.)

4 Department of Cardiology and Angiology I, Faculty of Medicine, Heart Center Freiburg University, University of Freiburg, Hugstetter Str. 55, 79106 Freiburg, Germany; Corinna.Nadine.Lang@uniklinik-freiburg.de

5 Department of Medicine III (Interdisciplinary Medical Intensive Care), Medical Center, Faculty of Medicine, University of Freiburg, Hugstetter Str. 55, 79106 Freiburg, Germany

* Correspondence: gernot.michael.lang@uniklinik-freiburg.de

† These authors contributed equally to this manuscript.

Citation: Saravi, B.; Li, Z.; Pfannkuche, J.; Wystrach, L.; Häckel, S.; E. Albers, C.; Grad, S.; Alini, M.; Richards, R.G.; Lang, C.; et al. Angiotensin II Type 1 Receptor Antagonist Losartan Inhibits TNF-α-Induced Inflammation and Degeneration Processes in Human Nucleus Pulposus Cells. *Appl. Sci.* **2021**, *11*, 417. https://doi.org/10.3390/app11010417

Received: 16 November 2020
Accepted: 30 December 2020
Published: 4 January 2021

Abstract: Our recent study detected the expression of a tissue renin–angiotensin system (tRAS) in human intervertebral discs (IVDs). The present study sought to investigate the impact of the angiotensin II receptor type 1 (AGTR1) antagonist losartan on human nucleus pulposus (NP) cell inflammation and degeneration induced by tumor necrosis factor-α (TNF-α). Human NP cells (4 donors; Pfirrmann grade 2–3; 30–37-years–old; male) were isolated and expanded. TNF-α (10 ng/mL) was used to induce inflammation and degeneration. We examined the impact of losartan supplementation and measured gene expression of tRAS, anabolic, catabolic, and inflammatory markers in NP cells after 24 and 72 h of exposure. T0070907, a PPAR gamma antagonist, was applied to examine the regulatory pathway of losartan. Losartan (1 mM) significantly impaired the TNF-α-induced increase of pro-inflammatory (nitric oxide and TNF-α), catabolic (matrix metalloproteinases), and tRAS (AGTR1a and angiotensin-converting enzyme) markers. Further, losartan maintained the NP cell phenotype by upregulating aggrecan and downregulating collagen type I expression. In summary, losartan showed anti-inflammatory, anti-catabolic, and positive phenotype-modulating effects on human NP cells. These results indicate that tRAS signaling plays an important role in IVD degeneration, and tRAS modulation with losartan could represent a novel therapeutic approach.

Keywords: intervertebral disc; renin–angiotensin system; degeneration; regeneration; spine; inflammation

1. Introduction

Low back pain (LBP) is one of the most common health problems and a leading cause of disability worldwide, resulting in an enormous socioeconomic burden [1]. Symptomatic intervertebral disc degeneration (IDD) is a major cause of LBP [2]. IDD is associated with extracellular matrix (ECM) degradation, the release of pro-inflammatory cytokines, alterations in spine biomechanics, increased angiogenesis, and neo-innervation, resulting in discogenic pain and disability [3,4]. Pro-inflammatory cytokines affect the production of relevant catabolic enzymes, leading to progressive ECM degeneration. Several studies have shown that pro-inflammatory markers, such as tumor necrosis factor-alpha (TNF-α) and interleukin (IL)-6, are increasingly expressed in degenerated human IVDs [4,5].

At present, none of the available treatment strategies address the actual underlying pathologic dysregulations leading to IDD. Affected patients in the early stages of IVD degeneration and not qualified for surgery do not benefit from conservative treatments [6]. Surgical strategies that include spinal fusion or total disc replacement (TDR) have been introduced and carried out to treat symptomatic patients, even though the long-term benefit is elusive compared with conservative treatment options [7–10]. Given the dearth of adequate self-repair capabilities of IVDs and satisfactory treatment approaches, new biomolecular therapies targeting the inflammatory and degenerative cycles are receiving attention [11]. These approaches attempt to reduce the inflammatory settings of degenerative and inflammatory IVDs in order to slow down the progressive pro-inflammatory cascade [12–15].

Recently, we showed that the tissue renin–angiotensin system (tRAS) is expressed in human IVDs, revealing a positive correlation with pro-inflammatory cytokines and catabolic enzymes [16]. Angiotensin II (AngII) is the major effector of the tRAS and is involved in inflammatory cell pathways. Angiotensinogen (ATG), its precursor protein, is cleaved by the protease renin to produce AngI, which is then converted to biologically active AngII by angiotensin-converting enzyme (ACE). RAS inhibitors are one of the most prescribed drugs globally [17]. Recent research has raised evidence that RAS inhibitors could reduce TNF-α production both in vitro and in vivo [18,19]. Price et al. showed in a rat model of rheumatoid arthritis (RA) that TNF-α production and pathological characteristics, such as knee joint swelling, were significantly diminished by ACE inhibitor application [20]. These findings were also confirmed by other groups, revealing that ACE inhibitors and angiotensin II receptor type 1 blockers (ARBs) are associated with anti-arthritic effects through the reduction of reactive oxygen species, inflammation, neutrophil recruitment, disease activity, and finally joint destruction [21–34]. Moreover, animal models of renal injury examining the protective impact of RAS inhibitors revealed that inhibition of AngII functions via ACE inhibitors or ARBs reduced the recruitment of pro-inflammatory cells and catabolic gene expression [35–37].

We hypothesize that a local RAS is involved in the progress of disc degeneration, inflammation, and discogenic pain in human IVD cells. As inflammation contributes to IDD, the present study sought to investigate whether inhibition of the angiotensin II receptor type 1 (AGTR1) would have a protective effect on IVD tissue biology. Specifically, we investigated the potential protective effects of angiotensin II receptor type 1 antagonist losartan on human nucleus pulposus (NP) cell inflammation and degeneration induced by TNF-α.

2. Materials and Methods

2.1. Isolation and Expansion of Human NP Cells

The Swiss Human Research Act does not apply to research that utilizes anonymized biological material and/or anonymously collected or anonymized health-related data. Therefore, this project does not need to be approved by an ethics committee. Patients' general consent, which also covers anonymization of health-related data and biological material, was obtained. Human NP waste tissue was collected from patients that underwent spinal surgeries with written consent (4 donors, male, 31–37 years old, Pfirrmann grade II). The collected NP tissue was incubated with red blood cell lysis buffer (155 mM NH_4Cl, 10 M $KHCO_3$, and 0.1 mM EDTA in Milli-Q water) for 5 min on a shaker at room temperature, and then washed with Phosphate-Buffered Saline (PBS). Chopped tissue was then enzymatically digested with 0.2% w/v pronase (Roche, Basel, Switzerland) in Alpha Minimum Essential Medium (αMEM, Gibco) for 1 h, followed by 65 U/mL collagenase type II (Worthington, Columbus, OH, USA) in αMEM/10% Fetal Calf Serum (FCS, Corning,

Corning, NY, USA) at 37 °C for 12–14 h. A single-cell suspension was obtained by filtering through a 100 μm cell strainer. Next, cells were seeded at a density of 10,000 cells/cm^2 and expanded with αMEM supplemented with 10% FCS, 100 U/mL penicillin, and 100 mg/mL streptomycin (1% P/S, Gibco). Cells were cultured at a hypoxic condition of 2% O_2. The culture medium was changed twice a week and cells were detached at approximately 80% confluence using a dissociation buffer composed of 0.05% trypsin/EDTA (Gibco) for 5 min at 37 °C. Cells were sub-cultured at a cell density of 3000 cells/cm^2 for expansion. Passage 1 NP cells were cryopreserved in liquid N_2. After thawing, NP cells were expanded with high-glucose Dulbecco's Minimum Essential Medium (high glucose DMEM, Sigma–Aldrich, St. Louis, MO, USA) and 10% FCS. Passage 3 NP cells were used in the present study.

2.2. Cytotoxicity Test of Losartan

Cytotoxicity studies were performed to test whether losartan induced cell death of human NP cells at different doses and time points. Human NP cells were seeded in 96-well plates at a density of 2000 cells per well for cell viability analysis. Two donors with three technical replicates per donor were used for analysis. The groups were exposed to DMEM with 1% ITS+, 1% non-essential amino acids (Sigma–Aldrich, St. Louis, MO, USA), and 50 μg/mL L-ascorbic acid 2-phosphate (Sigma–Aldrich, St. Louis, MO, USA). After 24 h, losartan (Tocris Bioscience, Bristol, UK) was added to the treatment groups at concentrations of 100, 250, 500, 750, and 1000 μM. After incubation for 24, 48, and 72 h, the cells were washed with PBS and then exposed to the Cell Titer Blue® reagent (Promega Corporation, Madison, WI, USA) diluted 1:5 in DMEM. Fluorescence intensity was determined with the Viktor 3 plate reader (Perkin Elmer, Waltham, MA, USA) after 4 h of incubation (ex/em 560/590 nm).

2.3. Effect of Losartan on Human NP Cells

Passage 3 NP cells were seeded in 12-well plates at a density of 20,000 cells/well. Overall, four donors with three replicates per donor were assessed in this experiment. Additionally, three wells per group and donor were assessed for DNA content. Cells were seeded in 2 mL DMEM with 10% FCS to allow cell attachment. On the day of seeding the 12-well plates (Day 0, baseline), samples were taken for normalization. After 24 h, the medium was changed to the experiment medium with six groups. The control group was cultured with 2 mL DMEM, 1% non-essential amino acids, 50 μg/mL L-ascorbic acid 2-phosphate, and 1% ITS+. We prepared high-dose (1000 μM) and low-dose (100 μM) stock solutions of losartan by dissolving 47.9 mg losartan potassium and 4.79 mg, respectively, in 1 mL PBS. After sterile filtering with a 0.2 μm filter, we prepared 250 μL aliquots and stored them at −20 °C. Losartan at concentrations of 100–1000 μM in PBS was then added to the medium for the experiments. Human recombinant TNF-α (10 ng/mL) was added simultaneously to induce a pro-inflammatory stimulus in human NP cells.

Experimental groups:

- Control group
- Losartan 100 μM
- Losartan 1000 μM
- TNF-α 10 ng/mL
- TNF-α 10 ng/mL + Losartan 100 μM
- TNF-α 10 ng/mL + Losartan 1000 μM

After treatment for 72 h, the medium was collected and analyzed for nitric oxide (NO) content. Three wells/group were digested with 0.5 mg/mL proteinase K (Sigma–Aldrich, St. Louis, MO, USA) to measure DNA content in the monolayer. Three wells/group were lysed in TRI reagent (Molecular Research Center, Cincinnati, OH, USA) and PolyAcryl-Carrier (Molecular Research Center, Cincinnati, OH, USA), and stored at −80 °C for gene expression analysis.

2.4. Pathway Study with the Peroxisome Proliferator-Activated Receptor Gamma (PPARγ) Antagonist T0070907 under Inflammatory Conditions

To determine a potential pathway for the anti-inflammatory effect of losartan, we conducted a pathway study with the PPARγ antagonist T0070907 (Tocris Bioscience, Bristol, UK). Human recombinant TNF-α (10 ng/mL) was used as an inducer of inflammation. T0070907 was added to the medium to investigate whether losartan's anti-inflammatory effect would be inhibited by blocking PPARγ. DMSO was used to solubilize T0070907 and the same amount (1 μL/mL medium) was also added to the other groups to provide equivalent culture conditions. As before, the cells were seeded on day 0 and exposed to the experimental group medium after 24 h. Gene expression analysis was performed after 24 and 48 h of exposure. The experimental groups were:

- Control
- TNF-α 10 ng/mL
- TNF-α 10 ng/mL + T0070907 1 μM
- TNF-α 10 ng/mL + Losartan 1 mM
- TNF-α 10 ng/mL + Losartan 1 mM + T0070907 1 μM

2.5. Gene Expression Analysis

Total RNA was extracted with 1-bromo-3-chloropropane (Sigma–Aldrich) followed by RNA precipitation using isopropanol and high salt precipitation solution (0.8 M sodium citrate and 1.2 M NaCl). RNA was washed with ethanol and quantified using a NanoDrop ND-1000 spectrophotometer (Thermo Fisher Scientific Inc., Waltham, MA, USA). SuperScript VILO cDNA Synthesis Kit (Life Technologies, Carlsbad, CA, USA) was used for reverse transcription, followed by real-time quantitative PCR (qRT-PCR) using the TaqMan™ method with 10 μL reaction volume.

qRT-PCR was performed to assess the gene expression of matrix metalloproteinase-1 and -3 (MMP-1, -3), interleukin 6 and 8 (IL-6, -8), TNF-α, and PPARγ as markers for inflammation and matrix degradation. Aggrecan (ACAN), collagen I, and collagen II (COL I, II) were analyzed as markers for ECM production and cell phenotype identification. The mRNA expression levels of the following components of tRAS were also quantified: angiotensin-converting enzyme (ACE), angiotensinogen (AGT), renin-like tRAS equivalent Cathepsin D, and the angiotensin II receptor type I (AGTR1). RPLP0 was used as a housekeeping gene in each sample (Table 1). The comparative CT method was applied for relative quantification with RPLP0 as the endogenous control.

2.6. Biochemical Analysis

DNA content was measured after overnight digestion with 0.5 mg/mL proteinase K (Sigma–Aldrich, St. Louis, MO, USA) at 56 °C. DNA quantification was performed with Hoechst 33258 (Sigma–Aldrich, St. Louis, MO, USA) dye and calf thymus DNA (Sigma–Aldrich, St. Louis, MO, USA) as the standard. Nitric oxide (NO), a marker for oxidative stress, was indirectly measured in the sampled medium of all wells by spectrophotometric quantification of nitrite, a non-volatile breakdown product, by Griess assay (Promega, Madison, WI, USA).

Table 1. Characteristics of custom-designed primer-probes and gene expression assays (Applied Biosystems) used for gene expression analysis.

Gene Acronym	Gene Full Name	Primer-Probe Sequence or Catalog Number	Reporter/Quencher
hRPLP0	Human 60S acidic ribosomal protein P0	Forward seq.: 5′-TGG GCA AGA ACA CCA TGA TG-3′ Reverse primer seq.: 5′CGG ATA TGA GGC AGC AGT TTC-3′	FAM/TAMRA
hACAN	Human Aggrecan	Forward seq.: 5′-AGT CTT CAA GCC TCC TGT ACT CA3′ Reverse primer seq.: 5′CGG GAA GTG GCG GTA ACA-3′	FAM/TAMRA
hACE	133 (48.90)	Hs01586213_m1	FAM/NFQ-MGB
hAGTR1a	Human angiotensin-II receptor type 1	Hs00258938_m1	FAM/NFQ-MGB
hCTSD	Human Cathepsin D	Hs00157205_m1	FAM/NFQ-MGB
hCOL1A1	Human collagen type 1 alpha 1 chain	Forward seq.: 5′-CCC TGG AAA GAA TGG AGA TGA T-3′ Reverse primer seq.: 5′ACT GAA ACC TCT GTG TCC CTT CA′3	FAM/TAMRA
hCOL2A1	Human collagen type 2 alpha 1 chain	Forward seq.: 5′-GGC AAT AGC AGG TTC ACG TAC A-3′ Reverse primer seq.: 5′GAT AAC AGT CTT GCC CCA CTT ACC-3′	FAM/TAMRA
hIL6	Human Interleukin 6	Hs00985639_m1	FAM/NFQ-MGB
hIL8	Human Interleukin 8	Hs00174103_m1	FAM/NFQ-MGB
hMMP1	Human matrix-metalloproteinase 1	Hs00899568_m1	FAM/NFQ-MGB
hMMP3	Human matrix-metalloproteinase 3	Hs00968305_m1	FAM/NFQ-MGB
hPPARγ	Human peroxisome proliferator-activated receptor gamma	Hs00234592_m1	FAM/NFQ-MGB
hTNFα	Human tumor necrosis factor α	Hs00174128_m1	FAM/NFQ-MGB

2.7. Enzyme-Linked Immunosorbent Assay

IL-6 content in the collected culture medium after 72 h of exposure was measured with an enzyme-linked immunosorbent (ELISA) kit (ELISA DuoSet®, R&D Systems, Catalog # DY008). The capture antibody was diluted to the working concentration and the microplate was coated with 100 μL per well with the diluted capture antibody, sealed, and incubated at room temperature overnight. On the next day, each well was aspirated and washed with a wash buffer (0.05% Tween20 in PBS, pH 7.2–7.4). The plates were then blocked by adding 300 μL of blocking buffer (1% Bovine Albumin Serum (BSA) in PBS, pH 7.2–7.4, filtered sterile (0.2 μm)) to each well. The plate was incubated at room temperature for a minimum of 1 h. The wash step was repeated and samples or standards were diluted in Reagent Diluent (0.1% Bovine Albumin Serum (BSA), 0.05% Tween 20 in Tris-buffered Saline (20 mM Trizma Base, 150 mM NaCl, pH 7.2–7.4, filtered sterile (0.2 μm)). Then, the plate was sealed and incubated for 2 h at room temperature. The wash step was repeated and 100μL of working dilution of Streptavidin-HRP was added to each well. The well was sealed and incubated at room temperature for 20 min and protected from light. The wash step was repeated and 100 μL of substrate solution (1:1 mixture of H_2O_2 and Tetramethylbenzidine) was added to each well. The plate was incubated for 20 min at room temperature and protected from light. Finally, the stop solution (2N H_2SO_4) was added to each well and the optical density of each well was determined using a microplate reader at 450 nm with wavelength correction at 570 nm.

2.8. Statistical Analysis

Statistical analysis was performed using GraphPad Prism 8.0 software (GraphPad Software, Inc., La Jolla, CA, USA) and Stata Statistical Software Release 15 (StataCorp. 2011, College Station, TX, USA). Data were assessed with the Shapiro–Wilk normality test. For normally distributed data, the differences were assessed using t-test or ANOVA, as appropriate. The Kruskal–Wallis test, followed by the Mann–Whitney U test for pair-wise comparisons, was used for non-parametric testing. A two-sided p-value < 0.05 was considered significant.

3. Results

3.1. Losartan Does Not Show Cytotoxic Effects on Human NP Cells

At concentrations between 100 and 1000 μM, losartan did not significantly affect the cell viability of human NP cells (Figure 1). We observed an increase in viable cell numbers from 24 to 72 h for the control group and all losartan groups examined. This increase in cell count was not different between different concentrations of losartan. These results indicate no cytotoxic effects of 100–1000 μM losartan on human NP cells after 24–72 h of exposure. We used losartan at a concentration of 100 μM and 1000 μM for subsequent experiments based on the cytotoxicity study results.

3.2. Effect of Losartan on Human NP Cell Proliferation and NO Release

The different experimental media did not significantly alter the DNA content of the human NP cells after exposure for 72 h. As expected, we observed a significant increase in NO release from the cells of the experimental groups containing TNF-α (p < 0.001). Losartan addition did not significantly affect NO release other than the TNF-α only group after 72 h of exposure (Figure 2).

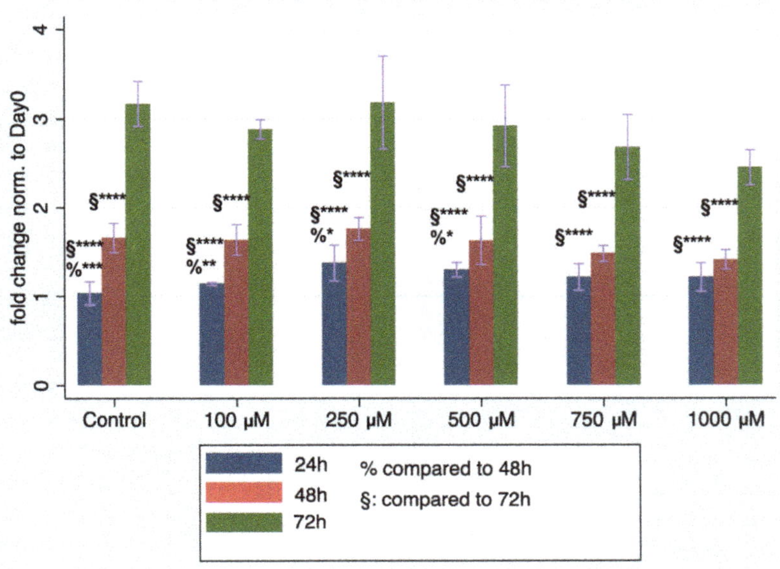

Figure 1. Relative viable cell count of human nucleus pulposus (NP) cells after 24, 48, and 72 h of losartan exposure. Means + standard deviations are shown. Data from two donors assessed in triplicate are shown ($n = 6$). %: compared to 48 h of exposure; §: compared to 72 h of exposure; * $p < 0.05$; ** $p < 0.01$; *** $p < 0.001$; **** $p < 0.0001$.

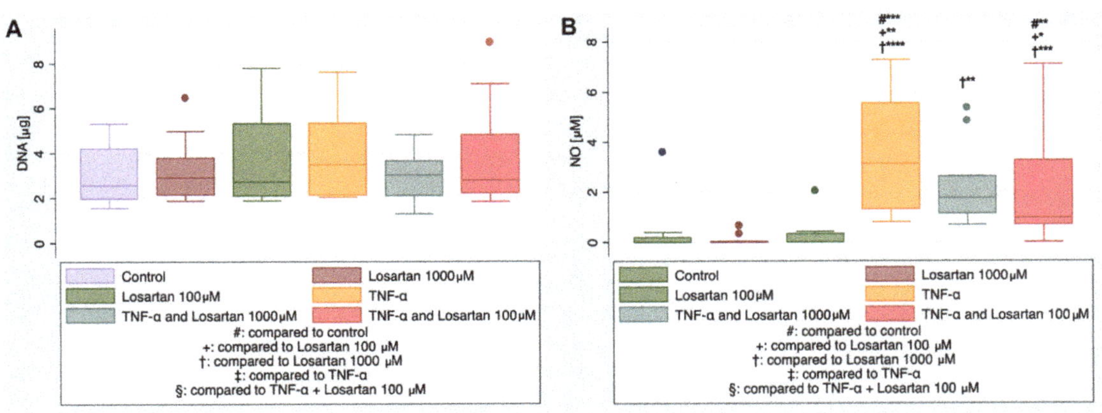

Figure 2. Effect of losartan and TNF-α on DNA content of human NP cells (**A**) and nitric oxide (NO) concentration in conditioned medium (**B**). Median and interquartile ranges (IQR) are shown. Dots represent outliers. Data from four donors assessed in triplicate are shown ($n = 12$). #: compared to control; +: compared to Losartan 100 µM; †: compared to Losartan 1000 µM; ‡: compared to TNF-α; §: compared to TNF-α + Losartan 100 µM; * $p < 0.05$; ** $p < 0.01$; *** $p < 0.001$; **** $p < 0.0001$.

3.3. Losartan Positively Modulates the Phenotype of Human NP Cells under Pro-Inflammatory Conditions

A significant upregulation of the pro-inflammatory cytokines IL-6 and IL-8 was observed under inflammatory conditions, which was unaltered by supplementation with losartan (Figure 3A). Gene expression of TNF-α was upregulated following the addition of TNF-α ($p < 0.01$), a trend that could be decreased by the addition of 1000 µM losartan ($p < 0.01$). Further, we found that the combination of 1000 µM losartan ($p < 0.001$)

and 100 μM losartan ($p < 0.01$) with TNF-α induced an increase in PPARγ gene expression compared with the control group, indicating that the angiotensin II receptor type 1 was involved in the regulation of PPARγ expression through AGTR1 in inflammatory settings. TNF-α significantly increased pro-inflammatory tRAS markers ACE ($p < 0.05$) and AGTR1 ($p < 0.01$). This effect was downregulated by blocking the AGTR1 receptor with 1000 μM losartan, which also downregulated AGT expression (TNF-α + Losartan vs. Control, $p < 0.01$), indicating that this upregulation of pro-inflammatory tRAS markers was mediated through the AGTR1 receptor under inflammatory situations (Figure 3B). TNF-α also upregulated the expression of Cathepsin D ($p < 0.01$), which was not altered by losartan addition. TNF-α induced upregulation of MMP-3 showed decreasing trends upon the addition of 1000 μM losartan, indicating an antidegenerative effect of high-dose losartan. However, this finding failed to reach significance ($p = 0.2676$) for the comparison TNF-α vs. TNF-α+ 1000 μM losartan. Interestingly, when TNF-α was present, the addition of losartan significantly upregulated the gene expression of ACAN (TNF-α+ 1000 μM losartan vs. TNF-α $p < 0.05$) and downregulated collagen I expression (TNF-α+ 1000 μM losartan vs. TNF-α, $p < 0.05$), indicating a positive phenotype-modifying effect of losartan under inflammatory conditions (Figure 3C). However, this observation was not significant for collagen II gene expression (TNF-α+ 1000 μM losartan vs. TNF-α, n.s.).

3.4. Pathway Study in Human NP Cells Treated with the PPARγ Antagonist T0070907 under Inflammatory Conditions

We found potential interactions between AGTR1 inhibition and PPARγ gene expression changes; hence, we evaluated whether losartan could directly interact with the PPARγ receptor. We examined the impact of an additional group containing the PPARγ antagonist T0070907 on respective genes after 24 h (Figure 4) and 72 h (Figure 5) of exposure. For the inflammatory markers IL-6 ($p < 0.01$) and TNF-α ($p < 0.05$), a significant upregulation of gene expression could be observed in the inflammatory groups compared with the control group after 24 h, confirming the findings of the previous experiments (Figure 3). In accordance with the previous experiments, this increase was not attenuated by losartan for IL-6. For TNF-α, a marked decrease was observed, which barely missed statistical significance (TNF-α vs. TNF-α + Losartan, $p = 0.0883$). The suppression of TNF-α-induced inflammation by the PPARγ antagonist T0070907 was not significant. A noticeable trend could be observed in the gene expression analysis of the catabolic marker MMP1. Here, significant upregulation was observed between the control group and inflammatory groups (Control vs. TNF-α, $p < 0.0001$), which could be decreased by the addition of losartan (TNF-α vs. TNF-α + Losartan, $p < 0.0001$). This effect was reversed with T0070907 in the medium (TNF-α + Losartan vs. TNF-α + Losartan + T0070907, $p < 0.0001$), indicating a partial agonism and anti-degenerative effects of losartan via the PPARγ receptor in human NP cells. Addition of TNF-α also led to a significant increase of ACAN. However, we did not observe any differences for the other experimental groups. Further, losartan reduced collagen I gene expression in TNF-α-treated NP cells (TNF-α vs. TNF-α + Losartan, $p < 0.001$). T0070907 intervention seemed to attenuate this effect of losartan, although the difference was not significant ($p = 0.1990$). For collagen II gene expression, we did not find any relevant significant changes.

Concerning the expression of the tRAS genes, we observed an upregulation of gene expression for AGTR1a, ACE, AGT, and Cathepsin D in inflammatory conditions (Control vs. TNF-α: ACE $p < 0.0001$, AGT $p < 0.0001$, AGTR1a $p < 0.01$; Figure 4). Losartan significantly reduced the upregulation of tRAS genes, which was reversed by the PPARγ antagonist T0070907 for ACE (TNF-α + Losartan vs. TNF-α + Losartan + T0070907 $p < 0.01$), Cathepsin D (TNF-α + Losartan vs. TNF-α + Losartan + T0070907 $p < 0.01$), and AGT (TNF-α + Losartan vs. TNF-α + Losartan + T0070907 $p < 0.001$; Figure 4). This trend seemed to be consistent for the angiotensin II receptor type 1, although not significant ($p = 0.4913$).

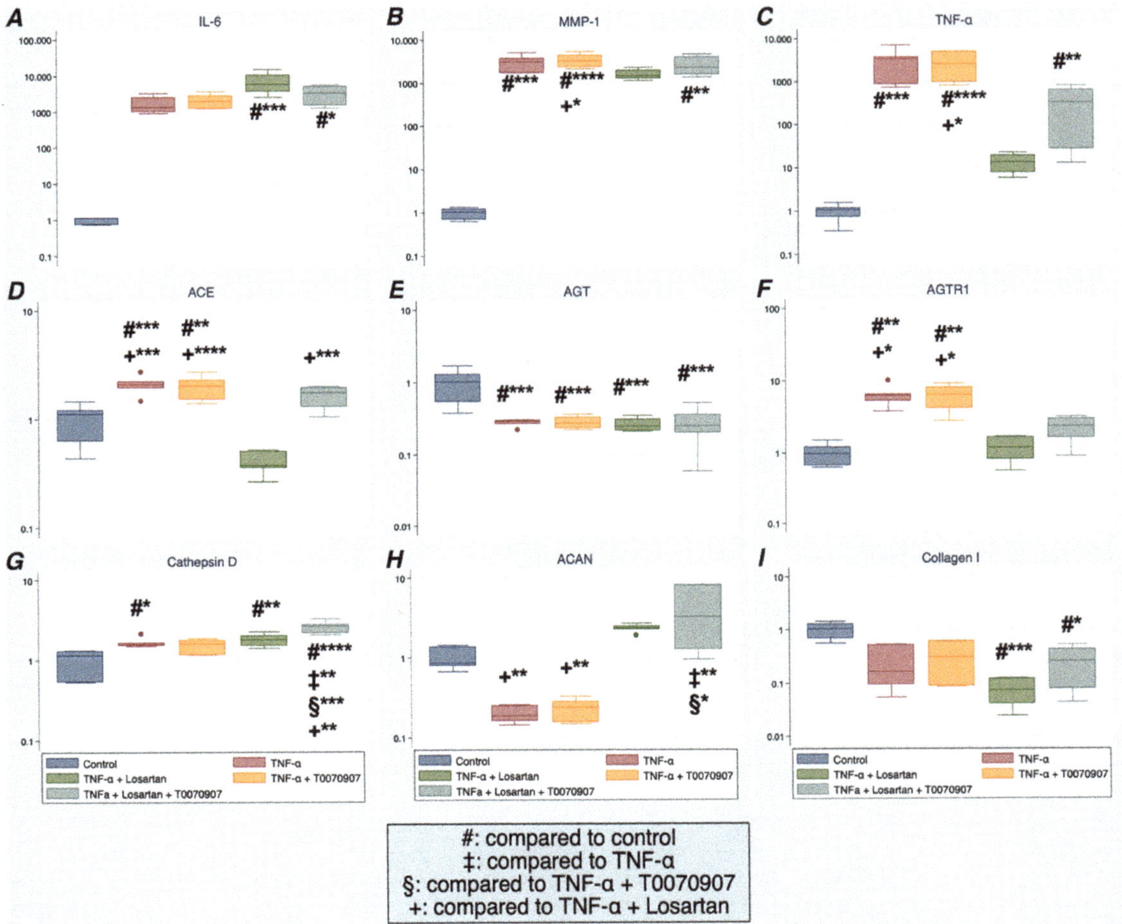

Figure 5. Influence of the PPARγ pathway on losartan-induced gene expression changes of inflammatory and tissue degeneration markers (**A–C**), tRAS markers (**D–G**), and cell phenotype markers (**H**,**I**) in human NP cells after 72 h of exposure. Gene expression was normalized to Day 0 (baseline) values. Data from two donors assessed in triplicate are shown ($n = 6$); * $p < 0.05$; ** $p < 0.01$; *** $p < 0.001$; **** $p < 0.0001$.

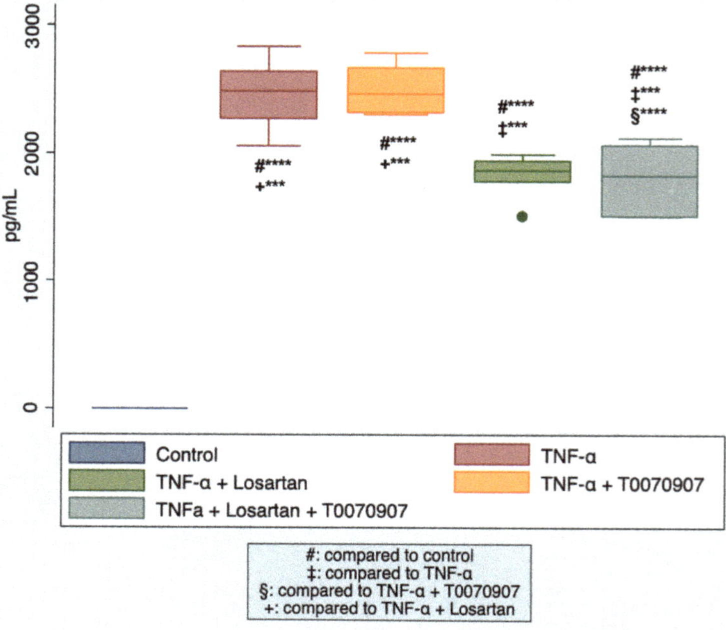

Figure 6. Influence of angiotensin II type 1 receptor (AGTR1) inhibition on secreted IL-6 levels in human NP cells after 72 h of exposure. The results are shown as the original concentrations in pg/mL in the media without normalization. Data from two donors assessed in triplicate are shown ($n = 6$); *** $p < 0.001$; **** $p < 0.0001$.

4. Discussion

The present study sought to investigate the protective effect of the AGTR1 antagonist losartan on human NP cell inflammation and degeneration induced by TNF-α. Outcomes revealed that TNF-α induced the expression of pathologic tRAS molecules and led to pro-inflammatory and catabolic effects in human NP cells. Inhibition of AGTR1 with losartan could partly inhibit the inflammatory and catabolic reaction. Therefore, present results suggest that TNF-α-induced disc degeneration may partially be mediated through AGTR1 signaling. Treatment with AGTR1 antagonist losartan could inhibit the TNF-α-induced degenerative state and maintain the NP cell phenotype depending on the administration or drug delivery method and the resulting achievable local tissue concentrations.

In accordance with previous studies, we found that losartan interacted with the PPARγ pathway [38]. Inhibition with the PPARγ inhibitor T0070907 partly abolished the effects of losartan on nucleus pulposus cells. This implies that the PPARγ pathway contributes to losartan's anti-inflammatory effects, as also suggested by Price et al. [20]. Our results also indicate that losartan has an impact on the gene expression of extracellular matrix-related components. Losartan has been reported to affect TGF-beta expression and reduce collagen I production in human fibroblasts [39]. Additionally, the interaction of losartan with TGF-beta signaling was confirmed by other research groups [40].

Targeting the tissue renin–angiotensin system may have significant therapeutic potential in modulating the metabolism of the degenerative IVD, which may potentially translate into a reduction of discogenic pain.

4.1. The Tissue Renin–Angiotensin System: AngII as a Pro-Inflammatory and Catabolic Hormone

The renin–angiotensin system (RAS) has an important role in the regulation and progression of tissue injuries in the cardiovascular system [41]. In earlier studies, a variety of locally acting renin–angiotensin system components were identified in various human tissues, such as bone, gastrointestinal tract, skin, kidney, and liver, revealing an important role in degenerative and inflammatory processes [42,43]. AngII, a well-known classical vasoconstrictory cardiovascular hormone and part of the pathological arm of the tRAS, was also demonstrated to be produced in inflammatory cells, inducing nerve growth and axon sprouting [44–48]. For the first time, Morimoto et al. and Price et al. revealed the existence of local renin–angiotensin systems and their contribution to inflammation in the musculoskeletal system of rats [20,49]. Recent evidence underlines the statement that the RAS and its main effector, AngII, may be considered as a locally acting system, regulating tissue homeostasis and regeneration, especially in the cardiovascular and nervous systems [50]. Furthermore, AngII induces the expression of pro-inflammatory markers, such as IL-6, TNF-α, and adhesion molecules, and functions as a true pro-inflammatory mediator that regulates inflammation, growth, and fibrosis [51–55].

In our previous work, we confirmed the existence of AngII and other tRAS components in tissue samples of degenerated discs by immunohistochemistry and gene expression analysis [16]. Disc tissue samples that expressed more tRAS factors showed significantly higher gene expression levels of pro-inflammatory (TNF-α IL-6) and catabolic genes (ADAMTS 4 and 5), indicating that tRAS contributes to the inflammatory processes operant in IDD. These findings are supported by the fact that the gene expression of NP phenotype-modulating factors, such as ACAN and COL2, was reduced in increased tRAS component-expressing discs. Furthermore, disc tissues with highly positive tRAS expression revealed lower glycosaminoglycan (GAG)/DNA ratios, implying the accelerated state of catabolism due to IDD.

Our present work indicates that AngII has the potential to affect IVD matrix degradation as the inhibition of its receptor, AGTR1, led to significant changes in the gene expression of relevant extracellular matrix genes. The roles of ACE and AGTR1 in inflammatory cell processes have previously been shown in synovium tissues from RA patients [20,56,57]. Recent research supports our work, revealing increased tRAS activity in the synovial fluid and tissues of patients with RA [58–60]. Price and coworkers analyzed the protective potential of losartan in rats with RA [20]. Chronic joint inflammation was induced by intraarticular and periarticular injection of Freund's complete adjuvant of heat-killed *Mycobacterium tuberculosis* into the knee joint. Acute joint inflammation was induced by intraarticular injection of λ-carrageenan and kaolin. Western blot analysis and immunohistochemistry reflected the elevated concentrations of AngII protein and AGTR1 in synovium from animals with acute and chronic joint inflammation. Losartan substantially reduced joint swelling and suppressed TNF-α generation in a dose-dependent manner. Morimoto et al. investigated the impact of AngII stimulation with different concentrations in rat annulus fibrosus cells by qRT-PCR [49]. Stimulation of rat IVD cells with AngII increased the mRNA expression of ADAMTS-5 significantly, indicating extracellular matrix degradation.

In summary, in accordance with other recent studies on musculoskeletal tissues, our current work provides strong evidence that the tRAS is involved in inflammatory and degenerative processes that are operant in IDD, and, therefore, introduces a novel therapeutic target to combat this devastating disease.

4.2. Anti-Inflammatory Therapies Via RAS Inhibition—A Potential Target in IDD?

Recent data consistently show that ARBs and ACE-inhibitors have anti-inflammatory effects in various human tissues and show beneficial effects in inflammatory musculoskeletal diseases [21–24,26–34,61]. Fukuzawa and coworkers demonstrated that oral administration of ACE inhibitors in mice reduced TNF-α release, though high concentrations were needed to reach a meaningful effect [18]. Further, RAS inhibitors abolished LPS-induced high IL-6 and TNF-α gene expression levels in the kidney [62]. Renal AngII is a key factor in mediating various components of the immune and inflammatory responses and acts as a pro-inflammatory agent [55,63]. This suggests that the administration of RAS inhibitor in therapeutic dosages to humans with hypertension may also suppress cytokine levels. The angiotensin-converting enzyme inhibitor Enalapril inhibits AngII synthesis from AngI and may suppress pro-inflammatory cytokine production, as previously shown in vitro [64,65]. Captopril, an ACE-inhibitor, is also known to have antirheumatic effects comparable to D-penicillamine [21,66,67]. Therefore, Captopril was considered a valuable drug in patients with hypertension and RA, especially due to the favorable benefit/risk ratio as Captopril lacks serious side effects [67]. Cardoso et al. demonstrated that losartan suppresses the expression of IL-6, IL-22, IL-17F, and IFN-γ in Peripheral Blood Mononuclear Cells (PBMCs) from RA patients, suggesting that losartan could be a superior option for hypertension treatment in RA patients [68].

Our results reflect previous findings that found downregulations of inflammatory markers in inflammatory cell culture models after AGTR1 blockade or knockdown [69,70]. In contrast to our experiments, these studies used lipopolysaccharides to stimulate inflammation in the respective cell cultures and might not be fully comparable. However, our workgroup recently validated the superior potential of TNF-α-induced inflammatory cell culture models to examine degenerative disc diseases [14]. Further, we showed for the first time that PPARγ pathway stimulation through losartan has protective effects on human NP cells. Losartan seems to exhibit at least some of its effect through the PPARγ pathway as the inhibition of PPARγ partly diminished the protective effects of losartan on the gene expression level. In accordance with the present results, several studies reported that losartan could get internalized through the AGTR1 receptor and act as a partial PPARγ agonist [69,71,72]. Noticeably, the PPARγ effects of ARBs might be too small regarding the reachable tissue levels to exert a significant anti-inflammatory effect. Therefore, new drug delivery methods and the development of new ARBs with more potent PPARγ activation properties or other PPARγ agonists might be needed if PPARγ stimulation is the target [73]. PPARγ activation properties of ARBs can be more seen as a beneficial effect in addition to the inhibition of the pathological tRAS arm (ACE/AngII/AGTR1 axis). Overall, more details about the interactions of the tRAS, PPARγ pathway, and ARBs in human IVD cells are warranted in the future.

Interestingly, the inhibition of AGTR1 receptors through losartan reduced TNF-α-induced ACE and AGTR1 upregulation. The upregulation of ACE through AGTR1 was already shown before by Koka et al. [74]. They also could inhibit this upregulation through losartan addition, which concurs with our data. Notably, as seen in our data, the inflammatory environments stimulate ACE and AGTR1 expression, indicating stimulation of the pathological arm of the tRAS in inflammatory settings. This was also shown by Takeshita et al., who suggested a previously unknown cross-talk between the TNF-α and the tRAS [75]. Therefore, the reproducible induction of the pathological arm of the tRAS by TNF-α for all examined donors in the present work suggests an important role in inflammatory and degenerative processes in IVDs. As shown by our data (Figure 3), losartan is especially effective in inflammatory settings, supporting its therapeutic potential in degenerative disc diseases.

4.3. Strengths and Limitations

The present study is associated with several strengths. The present experiments are the first pathway studies of the tRAS for IVD cells. As we used human NP cells from patients' IVDs in this study, these findings could directly impact clinical and therapeutic strategies in contrast to preclinical studies in other species that often need to be verified in humans first. We provided novel evidence that treatment with losartan suppresses pro-inflammatory and degenerative responses to inflammatory stimuli in human NP cells. We further revealed an interaction of the tRAS with the PPARγ pathway, which could be of potential interest for future pathway studies.

Some limitations need to be addressed to interpret the results adequately. Our studies were conducted in vitro using a 2-dimensional monolayer cell culture. In vivo interactions cannot be perfectly simulated with this methodological approach and the concentrations used for the intervention arms. The concentration range of losartan chosen in this experiment was based on previous in vitro pathway studies with other human cell types as evidence for intervertebral disc cells is scarce [69,76,77]. Therefore, the concentrations used in our cell cultures might not reflect the available tissue concentrations in humans after usual orally available doses of losartan. Reported oral doses of losartan range between 25 and 200 mg per day. Its terminal half-life is approximately 2 h, but its active metabolite EXP3174, which has a much more inhibitory effect on AGTR1 (up to 40-fold more potent), can reach a half-life of 6–9 h [78]. Reported maximal plasma concentrations in humans reached 84.5 ng/mL after a single oral administration of 25 mg and 1394.9 ng/mL after single oral administration of 200 mg losartan. For its active metabolite, these maximal plasma concentrations were 188.9 ng/mL and 2219.0 ng/mL, respectively [79]. This would suggest that oral doses cannot accumulate the tissue concentrations needed for the anti-inflammatory effects seen in our results, at least in vitro. New drug delivery and administration techniques will be needed to reach these tissue concentrations. Another approach would be to find more potent inhibitors of the pathological tRAS arm or to evaluate stimulation of the protective tRAS arm components, which possibly will lead to anti-inflammatory effects in much lower concentrations. However, the present preclinical experiments help us understand the role of the tRAS and its inhibition in human NP cells, and support future planning of potential ex-vivo organ culture models followed by the in vivo studies needed in order to confirm these results. Furthermore, the range of TNF-α concentrations used to establish an inflammatory environment in NP cells was based on our workgroup's previous works [15] and does not reflect the tissue concentrations of TNF-α found in humans, which is reported to be around 5.9–25.9 pg/mL depending on the duration of complaints [80]. As this is a preclinical experiment in 2D NP cells, the inflammatory environment shown in our in vitro model cannot exactly simulate the inflammatory situation in vivo. Future degenerative disc in vivo models are needed here to translate these preclinical findings. Notably, we focused on changes in gene expression levels, and these changes might not reflect the changes in protein levels. Our ongoing studies involving different tRAS modulators, such as other AGTR1 inhibitors, will be conducted using expanded methodological techniques in order to visualize and quantify the tRAS components in NP cells and quantify protein concentrations of important inflammatory and tissue remodeling markers released by the cells. This expansion of methological techniques is highly warranted to examine the pathways leading to the current results. Whereas we found a protective anti-inflammatory effect of losartan based on downregulation of secreted IL-6 into the culture medium, we could not observe an inhibition of the protective losartan effects after PPARγ inhibition. Therefore, we cannot make a final conclusion regarding the PPARγ interactions with the tRAS on the protein level at this time. Our current results that indicate an involvement of the PPARγ pathway are based on the gene expression changes and need further validation in the future. Furthermore, we observed differences in the outcome effects sizes regarding the two timepoints, indicating that genes might be differently affected by tRAS modulation depending on exposure time. For example, tRAS gene expressions (ACE, AGT, CTSD, AGTR1) seem to be more affected by 24 h of losartan exposure than 72 h. Future studies could include more time points and genes to ensure the identification of effects on relevant genes. Further, we did only observe significant effects on ACAN and collagen I gene expressions, but not collagen II gene expression, an important NP cell phenotype marker gene, indicating only partial effects on NP cell phenotype. Our ongoing experiments will include a broader examination of important genes and proteins that characterize the NP cell phenotype, in order to better evaluate the impact of tRAS modulation (including other tRAS modulators) on NP cell phenotype. Notably, we did not include other pivotal inflammatory markers such as prostaglandin E2, or cyclooxygenase-2. These markers need to be examined in the future, and this is already being planned by our workgroup, which will help clarify the interactions within NP cells. Moreover, other therapeutic approaches of tRAS modulation, such as the comparison of RAS inhibitors and stimulation of the protective tRAS arm, could be conducted and compared in order to evaluate the best therapeutic approach. Finally, an expansion of methodological techniques to quantify and visualize receptor expression changes is recommended and already planned by our group to characterize the nature of the tRAS in IVD degeneration.

5. Conclusions

TNF-α induced the expression of tRAS molecules and led to pro-inflammatory and catabolic effects in human NP cells. Inhibition of the angiotensin II receptor type 1 with losartan could inhibit this inflammatory and catabolic response. Further, we uncovered an interaction of losartan with the PPARγ pathway in human NP cells. These results demonstrate that TNF-α-induced disc degeneration may be mediated partially through AGTR1 signaling. Treatment with AGTR1 antagonist losartan could inhibit the TNF-α-induced degeneration process and maintain the NP cell phenotype. The inhibition of the pathological tRAS pathway with angiotensin II receptor type 1 blockers could be a novel therapeutic strategy for discogenic back pain caused by intervertebral disc degeneration.

Author Contributions: Conceptualization: Z.L., S.G., M.A., R.G.R., H.S., and G.L.; Data curation: B.S., Z.L., J.P., and L.W.; Formal analysis: B.S., Z.L., J.P., and L.W.; Investigation: S.G., M.A., C.L., N.S., H.S., and G.L.; Methodology: Z.L., J.P., C.E.A., S.H., S.G., and G.L.; Project administration: Z.L., S.G., M.A., R.G.R., N.S., H.S., and G.L.; Supervision: C.E.A., M.A., R.G.R., N.S., and G.L.; Validation: R.G.R. and C.L.; Visualization: B.S.; Writing—original draft: B.S. and G.L.; Writing—review & editing: B.S., Z.L., J.P., L.W., C.E.A., S.H., S.G., M.A., R.G.R., C.L., N.S., H.S., and G.L. All authors have read and agreed to the published version of the manuscript.

Funding: This study was funded by the German Spine Society (DWG), German Arthritis Foundation (DAH), the Foundation for the Promotion of Alternate and Complementary Methods to Reduce Animal Testing (SET) under the project InflamoDisc [number 59], AO Foundation, and AOSpine International. GL was supported by the Berta-Ottenstein Programme for Advanced Clinician Scientists, Faculty of Medicine, University of Freiburg.

Institutional Review Board Statement: Ethical approval was not necessary for this study. The Swiss Human Research Act does not apply to research that utilizes anonymized biological material and/or anonymously collected or anonymized health-related data. Therefore, this project does not need to be approved by an ethics committee.

Informed Consent Statement: Patients' informed consent was obtained from all subjects involved in the study.

Data Availability Statement: Datasets are available on request. The raw data and all related documents supporting the conclusions of this manuscript will be made available by the authors, without undue reservation, to any qualified researcher.

Acknowledgments: The article processing charge was funded by the Baden-Wuerttemberg Ministry of Science, Research and Art, and the University of Freiburg in the funding program Open Access Publishing.

Conflicts of Interest: The authors declare no conflict of interest.

References

1. Vos, T.; Abajobir, A.A.; Abate, K.H.; Abbafati, C.; Abbas, K.M.; Abd-Allah, F.; Abdulkader, R.S.; Abdulle, A.M.; Abebo, T.A.; Abera, S.F.; et al. Global, regional, and national incidence, prevalence, and years lived with disability for 328 diseases and injuries for 195 countries, 1990–2016: A systematic analysis for the Global Burden of Disease Study 2016. *Lancet* **2017**, *390*, 1211–1259. [CrossRef]
2. Battié, M.C.; Joshi, A.B.; Gibbons, L.E. Degenerative Disc Disease: What is in a Name? *Spine* **2019**, *44*, 1523–1529. [CrossRef] [PubMed]
3. Adams, M.A.; Roughley, P.J. What is Intervertebral Disc Degeneration, and What Causes It? *Spine* **2006**, *31*, 2151–2161. [CrossRef]
4. Risbud, M.V.; Shapiro, I.M. Role of cytokines in intervertebral disc degeneration: Pain and disc content. *Nat. Rev. Rheumatol.* **2014**, *10*, 44–56. [CrossRef] [PubMed]
5. Takahashi, H.; Suguro, T.; Okazima, Y.; Motegi, M.; Okada, Y.; Kakiuchi, T. Inflammatory Cytokines in the Herniated Disc of the Lumbar Spine. *Spine* **1996**, *21*, 218–224. [CrossRef]

6. Mannion, A.F.; Impellizzeri, F.M.; Leunig, M.; Jeszenszy, D.; Becker, H.-J.; Haschtmann, D.; Preiss, S.; Fekete, T.F. Eurospine 2017 full paper award: Time to remove our rose-tinted spectacles: A candid appraisal of the relative success of surgery in over 4500 patients with degenerative disorders of the lumbar spine, hip or knee. *Eur. Spine J.* **2018**, *27*, 778–788. [CrossRef]
7. Wahood, W.; Yolcu, Y.U.; Kerezoudis, P.; Goyal, A.; Alvi, M.A.; Freedman, B.A.; Bydon, M. Artificial Discs in Cervical Disc Replacement: A Meta-Analysis for Comparison of Long-Term Outcomes. *World Neurosurg.* **2020**, *134*, 598–613.e5. [CrossRef]
8. Gangl, M. Spezifischer Kreuzschmerz—Die erste Leitlinie. *Man. Med.* **2020**, *58*, 46–52. [CrossRef]
9. Findlay, C.; Ayis, S.; Demetriades, A.K. Total disc replacement versus anterior cervical discectomy and fusion: A systematic review with meta-analysis of data from a total of 3160 patients across 14 randomized controlled trials with both short- and medium- to long-term outcomes. *Bone Jt. J.* **2018**, *100-B*, 991–1001. [CrossRef] [PubMed]
10. Petersen, T.; Laslett, M.; Juhl, C. Clinical classification in low back pain: Best-evidence diagnostic rules based on systematic reviews. *BMC Musculoskelet Disord.* **2017**, *18*, 188. [CrossRef] [PubMed]
11. Rustenburg, C.M.E.; Faraj, S.S.A.; Ket, J.C.F.; Emanuel, K.S.; Smit, T.H. Prognostic factors in the progression of intervertebral disc degeneration: Which patient should be targeted with regenerative therapies? *JOR Spine* **2019**, *2*. [CrossRef] [PubMed]
12. Smith, L.J.; Silverman, L.; Sakai, D.; Le Maitre, C.L.; Mauck, R.L.; Malhotra, N.R.; Lotz, J.C.; Buckley, C.T. Advancing cell therapies for intervertebral disc regeneration from the lab to the clinic: Recommendations of the ORS spine section. *JOR Spine* **2018**, *1*, e1036. [CrossRef] [PubMed]
13. Buckley, C.T.; Hoyland, J.A.; Fujii, K.; Pandit, A.; Iatridis, J.C.; Grad, S. Critical aspects and challenges for intervertebral disc repair and regeneration-Harnessing advances in tissue engineering. *JOR Spine* **2018**, *1*, e1029. [CrossRef] [PubMed]
14. Li, Z.; Gehlen, Y.; Heizmann, F.; Grad, S.; Alini, M.; Richards, R.G.; Kubosch, D.; Südkamp, N.; Izadpanah, K.; Kubosch, E.J.; et al. Pre-clinical ex-vivo Testing of Anti-inflammatory Drugs in a Bovine Intervertebral Degenerative Disc Model. *Front. Bioeng. Biotechnol.* **2020**, *8*, 583. [CrossRef]
15. Du, J.; Pfannkuche, J.; Lang, G.; Häckel, S.; Creemers, L.B.; Alini, M.; Grad, S.; Li, Z. Proinflammatory intervertebral disc cell and organ culture models induced by tumor necrosis factor alpha. *JOR Spine* **2020**. [CrossRef]
16. Li, Z.; Wystrach, L.; Bernstein, A.; Grad, S.; Alini, M.; Richards, R.; Kubosch, D.; Südkamp, N.; Izadpanah, K.; Kubosch, E.; et al. The tissue-renin-angiotensin-system of the human intervertebral disc. *eCM* **2020**, *40*, 115–132. [CrossRef]
17. Stagnitti, M.N. Trends in Utilization and Expenditures of Prescribed Drugs Treating Diabetes, Hypertension, and High Cholesterol for Persons under Age 40 in the U.S. Civilian Noninstitutionalized Population, 2000 and 2010. In *Statistical Brief (Medical Expenditure Panel Survey (US))*; Agency for Healthcare Research and Quality (US): Rockville, MD, USA, 2001.
18. Fukuzawa, M.; Satoh, J.; Sagara, M.; Muto, G.; Muto, Y.; Nishimura, S.; Miyaguchi, S.; Qiang, X.L.; Sakata, Y.; Nakazawa, T.; et al. Angiotensin converting enzyme inhibitors suppress production of tumor necrosis factor-α in vitro and in vivo. *Immunopharmacology* **1997**, *36*, 49–55. [CrossRef]
19. Saravi, B.; Lang, G.; Ülkümen, S.; Burchard, T.; Weihrauch, V.; Patzelt, S.; Boeker, M.; Li, Z.; Woelber, J.P. The tissue renin-angiotensin system (tRAS) and the impact of its inhibition on inflammation and bone loss in the periodontal tissue. *Eur. Cell Mater.* **2020**, *40*, 203–226. [CrossRef]
20. Price, A.; Lockhart, J.C.; Ferrell, W.R.; Gsell, W.; McLean, S.; Sturrock, R.D. Angiotensin II type 1 receptor as a novel therapeutic target in rheumatoid arthritis: In vivo analyses in rodent models of arthritis and ex vivo analyses in human inflammatory synovitis. *Arthritis Rheum.* **2007**, *56*, 441–447. [CrossRef]
21. Agha, A.M.; Mansour, M. Effects of Captopril on Interleukin-6, Leukotriene B4, and Oxidative Stress Markers in Serum and Inflammatory Exudate of Arthritic Rats: Evidence of Antiinflammatory Activity. *Toxicol. Appl. Pharmacol.* **2000**, *168*, 123–130. [CrossRef]
22. Dalbeth, N. The non-thiol angiotensin-converting enzyme inhibitor quinapril suppresses inflammatory arthritis. *Rheumatology* **2005**, *44*, 24–31. [CrossRef] [PubMed]
23. Flammer, A.J.; Sudano, I.; Hermann, F.; Gay, S.; Forster, A.; Neidhart, M.; Künzler, P.; Enseleit, F.; Périat, D.; Hermann, M.; et al. Angiotensin-Converting Enzyme Inhibition Improves Vascular Function in Rheumatoid Arthritis. *Circulation* **2008**, *117*, 2262–2269. [CrossRef] [PubMed]
24. Liu, H.-M.; Wang, K.-J. Therapeutic effect of Captopril on rheumatoid arthritis in rats. *Asian Pac. J. Trop. Med.* **2014**, *7*, 996–999. [CrossRef]
25. Miyagi, M.; Ishikawa, T.; Kamoda, H.; Suzuki, M.; Murakami, K.; Shibayama, M.; Orita, S.; Eguchi, Y.; Arai, G.; Sakuma, Y.; et al. ISSLS Prize Winner: Disc Dynamic Compression in Rats Produces Long-Lasting Increases in Inflammatory Mediators in Discs and Induces Long-Lasting Nerve Injury and Regeneration of the Afferent Fibers Innervating Discs. *Spine* **2012**, *37*, 1810–1818. [CrossRef] [PubMed]
26. Tang, Y.; Hu, X.; Lu, X. Captopril, an angiotensin-converting enzyme inhibitor, possesses chondroprotective efficacy in a rat model of osteoarthritis through suppression local renin-angiotensin system. *Int. J. Clin. Exp. Med.* **2015**, *8*, 12584–12592.
27. Fahmy Wahba, M.G.; Shehata Messiha, B.A.; Abo-Saif, A.A. Ramipril and haloperidol as promising approaches in managing rheumatoid arthritis in rats. *Eur. J. Pharmacol.* **2015**, *765*, 307–315. [CrossRef]
28. Guerra, G.C.B.; de Menezes, M.S.S.; de Araújo, A.A.; de Araújo Júnior, R.F.; de Medeiros, C.A.C.X. Olmesartan Prevented Intra-articular Inflammation Induced by Zymosan in Rats. *Biol. Pharm. Bull.* **2016**, *39*, 1793–1801. [CrossRef]
29. Sagawa, K.; Nagatani, K.; Komagata, Y.; Yamamoto, K. Angiotensin receptor blockers suppress antigen-specific T cell responses and ameliorate collagen-induced arthritis in mice. *Arthritis Rheum.* **2005**, *52*, 1920–1928. [CrossRef]

30. Refaat, R.; Salama, M.; Abdel Meguid, E.; El Sarha, A.; Gowayed, M. Evaluation of the effect of losartan and methotrexate combined therapy in adjuvant-induced arthritis in rats. *Eur. J. Pharmacol.* **2013**, *698*, 421–428. [CrossRef]
31. Silveira, K.D.; Coelho, F.M.; Vieira, A.T.; Barroso, L.C.; Queiroz-Junior, C.M.; Costa, V.V.; Sousa, L.F.C.; Oliveira, M.L.; Bader, M.; Silva, T.A.; et al. Mechanisms of the anti-inflammatory actions of the angiotensin type 1 receptor antagonist losartan in experimental models of arthritis. *Peptides* **2013**, *46*, 53–63. [CrossRef]
32. Wang, D.; Hu, S.; Zhu, J.; Yuan, J.; Wu, J.; Zhou, A.; Wu, Y.; Zhao, W.; Huang, Q.; Chang, Y.; et al. Angiotensin II type 2 receptor correlates with therapeutic effects of losartan in rats with adjuvant-induced arthritis. *J. Cell. Mol. Med.* **2013**, *17*, 1577–1587. [CrossRef] [PubMed]
33. Queiroz-Junior, C.M.; Silveira, K.D.; de Oliveira, C.R.; Moura, A.P.; Madeira, M.F.M.; Soriani, F.M.; Ferreira, A.J.; Fukada, S.Y.; Teixeira, M.M.; Souza, D.G.; et al. Protective effects of the angiotensin type 1 receptor antagonist losartan in infection-induced and arthritis-associated alveolar bone loss. *J. Periodontal. Res.* **2015**, *50*, 814–823. [CrossRef] [PubMed]
34. Perry, M.E.; Chee, M.M.; Ferrell, W.R.; Lockhart, J.C.; Sturrock, R.D. Angiotensin receptor blockers reduce erythrocyte sedimentation rate levels in patients with rheumatoid arthritis. *Ann. Rheum. Dis.* **2008**, *67*, 1646–1647. [CrossRef] [PubMed]
35. Ruiz-Ortega, M.; Lorenzo, O.; Suzuki, Y.; Rupérez, M.; Egido, J. Proinflammatory actions of angiotensins. *Curr. Opin. Nephrol. Hypertens.* **2001**, *10*, 321–329. [CrossRef]
36. Wolf, G.; Neilson, E.G. Angiotensin II as a renal growth factor. *J. Am. Soc. Nephrol.* **1993**, *3*, 1531–1540.
37. Mezzano, S.A.; Ruiz-Ortega, M.; Egido, J. Angiotensin II and Renal Fibrosis. *Hypertension* **2001**, *38*, 635–638. [CrossRef]
38. Schupp, M.; Janke, J.; Clasen, R.; Unger, T.; Kintscher, U. Angiotensin Type 1 Receptor Blockers Induce Peroxisome Proliferator–Activated Receptor-γ Activity. *Circulation* **2004**, *109*, 2054–2057. [CrossRef]
39. Diop-Frimpong, B.; Chauhan, V.P.; Krane, S.; Boucher, Y.; Jain, R.K. Losartan inhibits collagen I synthesis and improves the distribution and efficacy of nanotherapeutics in tumors. *Proc. Natl. Acad. Sci. USA* **2011**, *108*, 2909–2914. [CrossRef]
40. Cohn, R.D.; van Erp, C.; Habashi, J.P.; Soleimani, A.A.; Klein, E.C.; Lisi, M.T.; Gamradt, M.; ap Rhys, C.M.; Holm, T.M.; Loeys, B.L.; et al. Angiotensin II type 1 receptor blockade attenuates TGF-β–induced failure of muscle regeneration in multiple myopathic states. *Nat. Med.* **2007**, *13*, 204–210. [CrossRef]
41. Namsolleck, P.; Recarti, C.; Foulquier, S.; Steckelings, U.M.; Unger, T. AT2 Receptor and Tissue Injury: Therapeutic Implications. *Curr. Hypertens. Rep.* **2014**, *16*, 416. [CrossRef]
42. Patil, J.; Schwab, A.; Nussberger, J.; Schaffner, T.; Saavedra, J.M.; Imboden, H. Intraneuronal angiotensinergic system in rat and human dorsal root ganglia. *Regul. Pept.* **2010**, *162*, 90–98. [CrossRef] [PubMed]
43. Paul, M.; Poyan Mehr, A.; Kreutz, R. Physiology of Local Renin-Angiotensin Systems. *Physiol. Rev.* **2006**, *86*, 747–803. [CrossRef] [PubMed]
44. Hoch, N.E.; Guzik, T.J.; Chen, W.; Deans, T.; Maalouf, S.A.; Gratze, P.; Weyand, C.; Harrison, D.G. Regulation of T-cell function by endogenously produced angiotensin II. *Am. J. Physiol. Regul. Integr. Comp. Physiol.* **2009**, *296*, R208–R216. [CrossRef] [PubMed]
45. Jurewicz, M.; McDermott, D.H.; Sechler, J.M.; Tinckam, K.; Takakura, A.; Carpenter, C.B.; Milford, E.; Abdi, R. Human T and Natural Killer Cells Possess a Functional Renin-Angiotensin System: Further Mechanisms of Angiotensin II–Induced Inflammation. *JASN* **2007**, *18*, 1093–1102. [CrossRef] [PubMed]
46. Chakrabarty, A.; Blacklock, A.; Svojanovsky, S.; Smith, P.G. Estrogen Elicits Dorsal Root Ganglion Axon Sprouting via a Renin-Angiotensin System. *Endocrinology* **2008**, *149*, 3452–3460. [CrossRef]
47. Côté, F.; Do, T.H.; Laflamme, L.; Gallo, J.-M.; Gallo-Payet, N. Activation of the AT $_2$ Receptor of Angiotensin II Induces Neurite Outgrowth and Cell Migration in Microexplant Cultures of the Cerebellum. *J. Biol. Chem.* **1999**, *274*, 31686–31692. [CrossRef]
48. Gendron, L.; Côté, F.; Payet, M.D.; Gallo-Payet, N. Nitric Oxide and Cyclic GMP Are Involved in Angiotensin II AT_2 Receptor Effects on Neurite Outgrowth in NG108-15 Cells. *Neuroendocrinology* **2002**, *75*, 70–81. [CrossRef]
49. Morimoto, R.; Akeda, K.; Iida, R.; Nishimura, A.; Tsujii, M.; Obata, S.; Kasai, Y.; Uchida, A.; Sudo, A. Tissue renin-angiotensin system in the intervertebral disc. *Spine* **2013**, *38*, E129–E136. [CrossRef]
50. Unger, T.; Steckelings, U.M.; dos Santos, R.S. *The Protective Arm of the Renin Angiotensin: Functional Aspects and Therapeutic Implications*; Academic Press: Cambridge, MA, USA, 2015; ISBN 978-0-12-801485-1.
51. Han, Y.; Runge, M.S.; Brasier, A.R. Angiotensin II Induces Interleukin-6 Transcription in Vascular Smooth Muscle Cells Through Pleiotropic Activation of Nuclear Factor-κB Transcription Factors. *Circ. Res.* **1999**, *84*, 695–703. [CrossRef]
52. Moriyama, T.; Fujibayashi, M.; Fujiwara, Y.; Kaneko, T.; Xia, C.; Imai, E.; Kamada, T.; Ando, A.; Ueda, N. Angiotensin II stimulates interleukin-6 release from cultured mouse mesangial cells. *J. Am. Soc. Nephrol.* **1995**, *6*, 95–101.
53. Ruiz-Ortega, M.; Ruperez, M.; Lorenzo, O.; Esteban, V.; Blanco, J.; Mezzano, S.; Egido, J. Angiotensin II regulates the synthesis of proinflammatory cytokines and chemokines in the kidney. *Kidney Int.* **2002**, *62*, S12–S22. [CrossRef] [PubMed]
54. Ekholm, M.; Kahan, T.; Jörneskog, G.; Bröijersen, A.; Wallén, N.H. Angiotensin II infusion in man is proinflammatory but has no short-term effects on thrombin generation in vivo. *Thromb. Res.* **2009**, *124*, 110–115. [CrossRef] [PubMed]
55. Ekholm, M.; Kahan, T.; Jörneskog, G.; Brinck, J.; Wallén, N.H. Haemostatic and inflammatory alterations in familial hypercholesterolaemia, and the impact of angiotensin II infusion. *J. Renin. Angiotensin. Aldosterone Syst.* **2015**, *16*, 328–338. [CrossRef] [PubMed]
56. Walsh, D.A.; Mapp, P.I.; Wharton, J.; Polak, J.M.; Blake, D.R. Neuropeptide degrading enzymes in normal and inflamed human synovium. *Am. J. Pathol.* **1993**, *142*, 1610–1621. [PubMed]

57. Walsh, D.A. Angiotensin converting enzyme in human synovium: Increased stromal [125I]351A binding in rheumatoid arthritis. *Ann. Rheum. Dis.* **2000**, *59*, 125–131. [CrossRef]
58. Goto, M.; Fujisawa, M.; Yamada, A.; Okabe, T.; Takaku, F.; Sasano, M.; Nishioka, K. Spontaneous release of angiotensin converting enzyme and interleukin 1 beta from peripheral blood monocytes from patients with rheumatoid arthritis under a serum free condition. *Ann. Rheum. Dis.* **1990**, *49*, 172–176. [CrossRef]
59. Goto, M.; Sasano, M.; Fuzisawa, M.; Okabe, T.; Nishizawa, K. Constitutive production of angiotensin converting enzyme from rheumatoid nodule cells under serum free conditions. *Ann. Rheum. Dis.* **1992**, *51*, 741–742. [CrossRef] [PubMed]
60. Veale, D.; Yanni, G.; Bresnihan, B.; FitzGerald, O. Production of angiotensin converting enzyme by rheumatoid synovial membrane. *Ann. Rheum. Dis.* **1992**, *51*, 476–480. [CrossRef] [PubMed]
61. Shi, Q.; Abusarah, J.; Baroudi, G.; Fernandes, J.C.; Fahmi, H.; Benderdour, M. Ramipril attenuates lipid peroxidation and cardiac fibrosis in an experimental model of rheumatoid arthritis. *Arthritis Res.* **2012**, *14*, R223. [CrossRef] [PubMed]
62. Niimi, R.; Nakamura, A.; Yanagawa, Y. Suppression of Endotoxin-Induced Renal Tumor Necrosis Factor-α and Interleukin-6 mRNA by Renin-Angiotensin System Inhibitors. *Jpn. J. Pharm.* **2002**, *88*, 139–145. [CrossRef] [PubMed]
63. Rüster, C.; Wolf, G. Angiotensin II as a Morphogenic Cytokine Stimulating Renal Fibrogenesis. *JASN* **2011**, *22*, 1189–1199. [CrossRef]
64. Schindler, R.; Dinarello, C.A.; Koch, K.-M. Angiotensin-converting-enzyme inhibitors suppress synthesis of tumour necrosis factor and interleukin 1 by human peripheral blood mononuclear cells. *Cytokine* **1995**, *7*, 526–533. [CrossRef]
65. Peeters, A.C.T.M.; Netea, M.G.; Kullberg, B.J.; Thien, T.; Van Der Meer, J.W.M. The effect of renin–angiotensin system inhibitors on pro- and anti-inflammatory cytokine production. *Immunology* **1998**, *94*, 376–379. [CrossRef]
66. Popa, C.D.; van Riel, P.L.C.M. The Use of Captopril in Rheumatoid Arthritis: Combining Treatment Targets! *Can. J. Cardiol.* **2013**, *29*, 639.e13. [CrossRef] [PubMed]
67. Martin, M.F.R.; Mckenna, F.; Bird, H.A.; Surrall, K.E.; Dixon, J.S.; Wright, V. CAPTOPRIL: A NEW TREATMENT FOR RHEUMATOID ARTHRITIS? *Lancet* **1984**, *323*, 1325–1328. [CrossRef]
68. Cardoso, P.R.G.; Matias, K.A.; Dantas, A.T.; Marques, C.D.L.; Pereira, M.C.; Duarte, A.L.B.P.; de Melo Rego, M.J.B.; da Rocha Pitta, I.; da Rocha Pitta, M.G. Losartan, but not Enalapril and Valsartan, Inhibits the Expression of IFN-γ, IL-6, IL-17F and IL-22 in PBMCs from Rheumatoid Arthritis Patients. *TORJ* **2018**, *12*, 160–170. [CrossRef] [PubMed]
69. An, J.; Nakajima, T.; Kuba, K.; Kimura, A. Losartan inhibits LPS-induced inflammatory signaling through a PPARγ-dependent mechanism in human THP-1 macrophages. *Hypertens. Res.* **2010**, *33*, 831–835. [CrossRef]
70. Wong, M.H.; Chapin, O.C.; Johnson, M.D. LPS-Stimulated Cytokine Production in Type I Cells Is Modulated by the Renin–Angiotensin System. *Am. J. Respir. Cell Mol. Biol.* **2012**, *46*, 641–650. [CrossRef]
71. Schupp, M.; Lee, L.D.; Frost, N.; Umbreen, S.; Schmidt, B.; Unger, T.; Kintscher, U. Regulation of Peroxisome Proliferator–Activated Receptor γ Activity by Losartan Metabolites. *Hypertension* **2006**, *47*, 586–589. [CrossRef]
72. Villar-Cheda, B.; Costa-Besada, M.A.; Valenzuela, R.; Perez-Costas, E.; Melendez-Ferro, M.; Labandeira-Garcia, J.L. The intracellular angiotensin system buffers deleterious effects of the extracellular paracrine system. *Cell Death Dis.* **2017**, *8*, e3044. [CrossRef]
73. Fujimura, A.; Ushijima, K.; Ando, H. Does the PPAR-γ-activating property of telmisartan provide a benefit in clinical practice? *Hypertens Res.* **2013**, *36*, 183. [CrossRef] [PubMed]
74. Koka, V.; Huang, X.R.; Chung, A.C.K.; Wang, W.; Truong, L.D.; Lan, H.Y. Angiotensin II Up-Regulates Angiotensin I-Converting Enzyme (ACE), but Down-Regulates ACE2 via the AT1-ERK/p38 MAP Kinase Pathway. *Am. J. Pathol.* **2008**, *172*, 1174–1183. [CrossRef] [PubMed]
75. Takeshita, Y.; Takamura, T.; Ando, H.; Hamaguchi, E.; Takazakura, A.; Matsuzawa-Nagata, N.; Kaneko, S. Cross talk of tumor necrosis factor-α and the renin–angiotensin system in tumor necrosis factor-α-induced plasminogen activator inhibitor-1 production from hepatocytes. *Eur. J. Pharmacol.* **2008**, *579*, 426–432. [CrossRef] [PubMed]
76. Soldner, A.; Benet, L.Z.; Mutschler, E.; Christians, U. Active transport of the angiotensin-II antagonist losartan and its main metabolite EXP 3174 across MDCK-MDR1 and Caco-2 cell monolayers: Active transport of losartan and EXP 3174. *Br. J. Pharmacol.* **2000**, *129*, 1235–1243. [CrossRef]
77. Chang, L.-T.; Sun, C.-K.; Chiang, C.-H.; Wu, C.-J.; Chua, S.; Yip, H.-K. Impact of simvastatin and losartan on antiinflammatory effect: In vitro study. *J. Cardiovasc. Pharm.* **2007**, *49*, 20–26. [CrossRef]
78. Sica, D.A.; Gehr, T.W.B.; Ghosh, S. Clinical pharmacokinetics of losartan. *Clin. Pharm.* **2005**, *44*, 797–814. [CrossRef]
79. Ohtawa, M.; Takayama, F.; Saitoh, K.; Yoshinaga, T.; Nakashima, M. Pharmacokinetics and biochemical efficacy after single and multiple oral administration of losartan, an orally active nonpeptide angiotensin II receptor antagonist, in humans. *Br. J. Clin. Pharm.* **1993**, *35*, 290–297. [CrossRef]
80. Altun, I. Cytokine profile in degenerated painful intervertebral disc: Variability with respect to duration of symptoms and type of disease. *Spine J.* **2016**, *16*, 857–861. [CrossRef]

Article

Screening for Growth-Factor Combinations Enabling Synergistic Differentiation of Human MSC to Nucleus Pulposus Cell-Like Cells

Kosuke Morita [1,†], Jordy Schol [1,†], Tibo N. E. Volleman [2], Daisuke Sakai [1,3,*], Masato Sato [1,3] and Masahiko Watanabe [1,3]

1 Department of Orthopaedic Surgery, Tokai University School of Medicine, 143 Shimokasuya, Isehara, Kanagawa 259-1193, Japan; km6amm1112@yahoo.co.jp (K.M.); schol.j@tsc.u-tokai.ac.jp (J.S.); sato-m@is.icc.u-tokai.ac.jp (M.S.); masahiko@is.icc.u-tokai.ac.jp (M.W.)
2 Department of Biomedical Engineering, Eindhoven University of Technology, 5612AZ Eindhoven, The Netherlands; tibovolleman@gmail.com
3 Center for Musculoskeletal Innovative Research and Advancement (C-MiRA), Tokai University Graduate School, 143 Shimokasuya, Isehara, Kanagawa 259-1193, Japan
* Correspondence: daisakai@is.icc.u-tokai.ac.jp
† Equally contributing authors.

Abstract: Background: Multiple studies have examined the potential of growth factors (GF) to enable mesenchymal stromal cells (MSC) to nucleus pulposus (NP) cell-like cell differentiation. Here we screened a wide range of GF and GF combinations for supporting NP cell-like cell differentiation. **Methods:** Human MSC were stimulated using 86 different GF combinations of TGF-β1, -2, -3, GDF5, -6, Wnt3a, -5a, -11, and Shh. Differentiation potency was assessed by alcian blue assay and NP cell marker expression (e.g., COL2A1, CD24, etc.). The top four combinations and GDF5/TGF-β1 were further analyzed in 3D pellet cultures, on their ability to similarly induce NP cell differentiation. **Results:** Almost all 86 GF combinations showed their ability to enhance proteoglycan production in alcian blue assay. Subsequent qPCR analysis revealed TGF-β2/Wnt3a, TGF-β1/Wnt3a, TGF-β1/Wnt3a/GDF6, and Wnt3a/GDF6 as the most potent combinations. Although in pellet cultures, these combinations supported NP marker expression, none showed the ability to significantly induce chondrogenic NP matrix production. Only GDF5/TGF-β1 resulted in chondrogenic pellets with significantly enhanced glycosaminoglycan content. **Conclusion:** GDF5/TGF-β1 was suggested as an optimal GF combination for MSC to NP cell induction, although further assessment using a 3D and in vivo environment is required. Wnt3a proved promising for monolayer-based NP cell differentiation, although further validation is required.

Keywords: nucleus pulposus; growth factors; differentiation; mesenchymal stromal cell; growth factor; Wnt; differentiation factor; chondrogenesis; cell therapy

Citation: Morita, K.; Schol, J.; Volleman, T.N.E.; Sakai, D.; Sato, M.; Watanabe, M. Screening for Growth-Factor Combinations Enabling Synergistic Differentiation of Human MSC to Nucleus Pulposus Cell-Like Cells. *Appl. Sci.* **2021**, *11*, 3673. https://doi.org/10.3390/app11083673

Academic Editor: Antonio Scarano

Received: 28 February 2021
Accepted: 15 April 2021
Published: 19 April 2021

1. Introduction

The primary causes of disability worldwide are low back and neck pain [1]. Despite their prevalence and socioeconomic impact, long-term curative treatments for these disorders are lacking. Novel approaches are being explored to resolve this issue, mainly focusing on alleviation or reversal of intervertebral disc (IVD) degeneration [2–4]. IVD degeneration is a progressive pathology, involving a decrease in cell numbers and cell potency, primarily within the central nucleus pulposus (NP) of the IVD, resulting in a switch in extracellular matrix (ECM) production [5,6]. Specifically, the initially proteoglycan-rich and type II collagen NP-ECM progressively turns into a type I collagen fibrotic tissue, undermining the IVD's biomechanical characteristics [7]. The responding loss in IVD integrity alters disc height, requiring compensation in other spinal components, thereby potentially leading to a variety of spinal disorders such as facet joint arthritis and spondylolisthesis. Alternatively, the outer collagen layers, collectively termed the annulus fibrosus (AF), can similarly lose their integrity, thereby enabling disc bulging or complete AF rupture. Finally, in reaction to these degenerative changes, the native IVD cells secrete inflammatory factors, potentially stimulating nerve ingrowth, vascularization, and immunogenic cell influx in the normally avascular and un-innervated discs or sensitizing nearby nerve structures [8–10]. Collectively, these changes can trigger nociception and in severe cases, might result in debilitating pain and disability in patients [11,12].

This vicious cycle of disc degeneration [6] has proven difficult to combat in a clinical setting. Conservative treatment generally involves pain medication or muscle relaxants, while surgical intervention at later stages of degeneration, involves removal of disc tissue (e.g., microdiscectomy) or replacement of the complete disc (e.g., arthrodesis or arthroplasty). These final stage interventions are costly, invasive, and come with a range of complications such as adjacent disc disease, surgical infections, etc. [13,14]. Moreover, their effectiveness in alleviating pain also remains to be confirmed [15]. Moreover, a large treatment gap exists for disc degeneration associated with LBP in the mild to moderate range, for patients for which analgesics proved ineffective but are not yet indicated for surgery.

The field of regenerative medicine promises a wide range of therapeutics that have been theorized to be effective against disc degeneration. A variety of therapeutic strategies are being explored, each aimed to support cells to regenerate the IVD, e.g., gene therapy [16], growth factor injection [17], biomaterial injection [18,19], or tissue engineering [20,21]. Farthest in its clinical development is cell therapy, a regenerative strategy in which *de novo* cells are introduced directly into the IVD [2,3]. There, transplanted cells can either integrate within the IVD and contribute to appropriate ECM production, or otherwise modulate endemic cells to change the catabolic/inflammatory state to a more anabolic one. Multiple animal studies [22] have suggested the ability to modulate disc degeneration and induce regeneration of its matrix composition [23,24] and inflammatory environment [25]. In the last decade, a multitude of clinical trials reported initial results, confirming the safety of cell therapy and have suggested their ability to reduce pain and disability [2]. However, large-scale placebo-controlled trials are still highly anticipated [2].

The primary cell source being investigated for IVD regeneration are MSC [26]. MSC are advantageous due to their accessibility, expandability, and suggested limited immunogenicity [27]. These, in contrast to NP cells or other chondrogenic cells, which lose their potency during culture [28] and have limited and compromised tissue sources that present low cell yields [29]. Furthermore, MSC possess immunomodulatory features, as well as multipotent differentiation capacity, and could thus potentially target IVD degeneration from two approaches. Nevertheless, the hypoxic and acidic IVD environment is foreign to MSC and questions remain about their ability to survive and thrive in such a demanding niche [30,31]. A recent report from the Mesoblast phase II/III clinical trial suggested their MSC-based product was able to alleviate some pain-associated outcomes, but no regenerative changes were observed on MRI modalities [32]. As such, a range of studies has attempted to induce the differentiation of MSC toward an NP cell-like cell phenotype. MSC can be subjected to environmental conditions [33], hydrogel systems [34,35], genetic manipulation [36,37], etc.; aimed to induce more NP cell-like features. Particularly, the application of growth and differentiation factors to support differentiation towards an NP cell-like cell, as indicated by enhanced proteoglycan production rates or enhanced NP cell marker expression levels; e.g., aggrecan, type II collagen, Brachyury, and CD24 [38]. Multiple growth factors are suggested to be able to support NP cell differentiation, including Wnt3a [39], Wnt5a [40], TGF-β1 [41], TGF-β3 [42], GDF5 [43], and GDF6 [44]. Very few bodies of work have compared different growth factor potency for NP cell differentiation. One study by Clarke et al. [44] suggested GDF6 media supplementation optimal for MSC to NP cell differentiation, following assessment of GDF5, GDF6, or TGF-β1 media supplementation. Moreover, work by Hodgkinson et al. revealed quite eloquently, the potential of GDF6 to differentiate adipose stem cells toward NP cell-like cells [45], and subsequently highlighted a requirement of the SMAD1/5/8 pathway [46]. Alternatively, work from Colombier et al. [47] revealed the ability of growth factors to synergistically enhance NP cell differentiation. In their work, they found that only combined application of TGF-β1 and GDF5 was able to create MSC that formed fully chondrogenic pellets, as well as stringently enhance expression of NP cell markers; e.g., ACAN, COL2A1, CD24, PAX1, and OVOS2. Review work has also suggested the application of both Wnt3a or Wnt5a as promising chondrogenic factors, particularly combining Wnt3a with other chondrogenic factors was speculated to hold potential [48]. A wide range of factors has been proposed to enable MSC to NP cell differentiation. However, a large-scale assessment or screening to determine which factors, or even more, which combination of factors, are most potent to induce NP cell features, has not been reported. To optimize growth factor-mediated MSC differentiation, particularly for the creation of cell therapy products, we initiated this study to screen a wide range of growth factor combinations to assess which combination showed most potent in inducing proteoglycan production and NP cell-associated expression levels as indicators of MSC to NP cell-like cell differentiation.

2. Materials and Methods

Study protocols and design, which involve the collection and application of human tissue samples, were reviewed and approved by the institutional ethics review committee of the Tokai University School of Medicine. (application number No. 06I-49 and 16R-051). Informed consent was obtained for the collection and usage of tissue materials from the respective patients.

2.1. Cell Isolation and Culture

As part of adult spinal fusion surgery procedures, bone marrow cells were collected and isolated using gradient centrifugation as previously described [49]. The cells were collected and cryopreserved in a commercial cryopreservation medium (Cellbanker®-1, Nippon Zenyaku Kogyo Co., Ltd., Koriyama, Japan) until further usage. At the start of the experiment, the mononuclear cells were rapidly thawed, gently washed, and seeded in αMEM (Wako Pure Chemical Industries, Ltd., Osaka, Japan) 20% (*v*/*v*) FBS (Sigma-Aldrich, St. Louis, MO, USA), 1% (*v*/*v*) Penicillin/Streptomycin (P/S; 10,000 U/mL; Gibco, Gaithersburg, MD, USA) and cultured for 2 weeks uninterrupted at approximately 21% O_2, 5% CO_2 at 37 °C. The unattached cells were aspirated, and media was refreshed. Alternatively, MSC were obtained commercially from LONZA Group AG (Basel, Switzerland) and similarly cultured. MSC were further expanded up to a confluency of ~80% upon which 0.25% Trypsin/EDTA (Thermo-Fisher, Waltham, MA, USA) mediated passaging was performed.

Additionally, human NP cells were obtained from surgical tissue material during lumbar disc herniation microdiscectomy. NP cells were isolated as previously described [29]. In brief, IVD tissue was collected in saline and macroscopically glossy NP tissue was separated. The tissues were fragmented by a scalpel and digested in Tryple Express (Thermo-Fisher) for 1 h. The subsequent suspension was further digested in collagenase P solution. The resulting cell suspension was filtered and cultured at 2–5% O_2, 5% CO_2 within 10% (*v*/*v*) FBS, 1% (*v*/*v*) P/S, αMEM (Thermo-Fisher).

2.2. Growth Factor-Induced Differentiation Screening

Differentiation assays in monolayer cultures were performed in Dulbecco's Modified Eagle Medium (DMEM; Wako Pure Chemical Industries) containing 10% (*v*/*v*) FBS and 1% (*v*/*v*) P/S, 50 nM Ascorbic Acid (Wako Pure Chemical Industries), 1% (*v*/*v*) Insulin, Transferrin, Selenium, Ethanolamine Solution (ITS-X; Beckton Dickinson, Franklin Lakes, NJ, USA), and 10 nM dexamethasone (Sigma-Aldrich), further referred to as differentiation medium [47]. Differentiation medium was further supplemented with different combinations of TGFβ-1 (Peprotech, Rocky Hill, NJ, USA), TGFβ-2 (Peprotech), and TGFβ-3 (Peprotech) at concentrations of 10 ng mL^{-1} following the work of Colombier et al. [47]

and Clarke et al. [44], GDF5 (Peprotech) and GDF6 (Peprotech) at concentrations of 100 ng mL^{-1} following the work of Colombier et al. [47] and Clarke et al. [44], Wnt3a (Peprotech) at a concentration of 250 ng mL^{-1}, Wnt5a (Peprotech) at a concentration of 50 ng mL^{-1} based on the work of Gibson et al. [50], Wnt11 (Peprotech) at a concentration of 50 ng mL^{-1}, and Sonic hedgehog (Shh; Peprotech) at a concentration of 100 ng mL^{-1}. Maximally three growth factors were combined. A negative sample, involving identically cultured MSC without growth factor supplementation was included. The induced MSC were cultured under 2% O_2 at 37 °C culture environment, with media replenishment every 2–3 days.

2.3. Alcian Blue Assay for Quantification of Glycosaminoglycan Content

Alcian blue assay was performed in triplicate by seeding unstimulated MSC (22-year-old) at a density of 6000 cells/cm^2 on a 96-wells plate and cultured with differentiation media supplemented with indicated growth factor combinations for 3 weeks. Subsequently, the cells were fixed in 10% formaldehyde (Wako Pure Chemical Industries) for 15 min at room temperature. Cells were thoroughly washed and subjected to 50 µL 1% (*w*/*v*) alcian blue (MERCK & Co., Kenilworth, NJ, USA), 0.1 M HCL for 60 min. Stained samples were thoroughly rinsed and digested overnight by 150 µL 6 M Guanidine (WAKO) HCL. The resulting samples were analyzed by measurement of absorbance at 650 nm using SpectraMax® i3 (Molecular Devices®, San Jose, CA, USA). Absorbance values were calculated relative to the negative control of MSC cultured without growth factors.

2.4. Chondrogenic Pellet Cultures

For chondrogenic pellet cultures, MSC (n = 2; 23- and 31-year-old) were seeded at a density of 3000 cells/cm^2 in T-75 culture flasks (IWAKI, Japan) with αMEM containing 20% FBS, 1% P/S. Upon 70–80% confluency, MSC were harvested and aliquoted at 2.5×10^5 cells per 15 mL polypropylene falcon tubes (Corning Inc., Midland, NC, USA). Chondrogenic pellet culture was initiated by spinning down the tubes at 500 g for 5 min. The resulting cellular pellet was cultured for 1 day at 21% O_2 at 37 °C. The next day, the medium was replaced by differentiation media as previously described. Based on the alcian blue assay the growth factor combinations; GDF5/TGF-β1, TGF-β2/Wnt3a, TGF-β1/Wnt3a, TGF-β1/Wnt3a/GDF6, GDF6/Wnt3a were applied. As a negative control, unstimulated MSC were identically cultured but were not stimulated with growth factors. The pellets were incubated with media replenishment thrice a week, at 37 °C under 5% CO_2 and hypoxic conditions of 2% O_2. Additionally, human NP cells obtained from surgical waste material were similarly cultured in differentiation media without growth factors (n = 3).

2.5. Wnt3a Mediated Pre-Conditioning

A recent literature review [48] suggested the potential ability of Wnt3a to maintain the differentiation potency of MSC as opposed to inducing chondrogenic differentiation, which could help explain the discrepancies in monolayer and pellet culture outcomes. To test this hypothesis, we applied Wnt3a at 250 ng mL^{-1} during the MSC (n = 1, 23-year-old) expansion culture and subsequently subjected these cells to chondrogenic pellet culture supplemented with GDF5/TGF-β1 or commercial pre-defined Mesenchymal Stem Cell Chondrogenic Differentiation Medium (Chon.M.: PromoCell, Heidelberg, Germany).

2.6. PicoGreen Assay for Quantification of DNA Content

Following 3 weeks of chondrogenic pellet culture, cells were washed and subjected to overnight papain digestion solution (125 µg mL^{-1} Papain-enzyme (Sigma-Aldrich), 8 mg mL^{-1} sodium acetate (Wako Pure Chemical Industries), 4 mg mL^{-1} Ethylene diamine tetra acetic acid disodium salt (EDTA2Na; Sigma-Aldrich), 0.8 mg mL^{-1} L-Cysteine (Sigma-Aldrich) 0.2 sodium phosphate buffer (pH 6.4)) at 65 °C. The resulting suspension was centrifuged for 10 min at 8400 rpm, after which the supernatant was collected and stored at −80 °C until further use. Thereafter, a fraction of the suspensions were analyzed using a QUANT-IT PicoGreen Kit (ThermoFisher) following the manufacturer's instruction. Briefly, the suspension was transferred to a 96-well assay microplate (black, transparent bottom; Corning Inc.). 100 µL of PicoGreen solution was added to each sample, and agitated in an environment devoid of light, for 3 min. 520 nm fluorescence emission was measured under 480 nm excitation using a SpectraMax® i3. DNA concentrations were determined by interpolating the fluorescence values from a standard curve obtained by adding solutions with known concentrations of DNA.

2.7. DMMB Assay for Glycosaminoglycan Content

Glycosaminoglycan content was measured by 1,9-dimethyl methylene blue (DMMB) staining, using the Glycosaminoglycan assay Blycan™ kit (Biocolor, Carrickfergus, UK) following the manufacturer's instructions. In short, the papain digested solution was mixed with Blycan DMMB and incubated for 30 min at room temperature. Samples were spun down for 10 min. The supernatant was discarded, and 0.5 mL dissociation reagent was added followed by 2-h incubation. The resulting suspension was centrifuged for 10 min and 150 µL of each sample was transferred to a 96-wells plate. Absorbance values were measured at 656 nm using the SpectraMax® i3 and concentration values were interpolated from a standard curve obtained from solutions with known concentrations of glycosaminoglycans.

2.8. RNA Isolation and Gene Expression Analysis

Monolayer cultures (n = 2; 23- and 31-year-old) were processed using SV total RNA System (Promega, Madison, WI, USA) following the manufacturer's instruction. DNAse treatment was employed to digest genomic DNA. Chondrogenic 3D pellet cultures were harvested for RNA isolation, by digestion in 1 mL Trizol (ThermoFisher) and subsequent grinding using RNase free stainless-steel balls and 3 min 3000 rpm shakes using a Shake Master NEO (Hirata, Kita, Japan). Subsequent suspensions were processed by Qiagen RNeasy kit (Qiagen, Hilden, Germany) according to the manufacturer's instruction to isolate the respective RNA molecules.

Quality and quantity of RNA were determined via NanoDropTM Lite spectrophotometer (Thermo Scientific) Subsequently, RNA was reverse transcribed into cDNA using a High-Capacity RNA-to-cDNA kit (Applied Biosystems, Beverly, MA, USA). Relative quantification of mRNA was performed using an SYBR Green Master Mix approach via QuantStudio3 (Applied Biosystems) employing custom primer sequences (Table A1). qPCR analysis of each sample was performed in duplicates. Obtained C_T values were compared via $2^{-\Delta\Delta CT}$ analysis, calculating gene expressions level relative to the housekeeping gene GAPDH and the average ΔC_T value of the growth-factor unstimulated MSC.

2.9. Tissue Processing and Histology

Chondrogenic pellets were washed and fixed using 10% formalin solution for 15 min. Samples were thereafter processed by incubating them for 1 h in subsequent 10% (w/v), 20% (w/v), and 30% (w/v) sucrose solutions. Finally, the pellet samples were submerged in an OCT TEK compound (Sakura Finetek, Torrance, USA) and snap-frozen in liquid nitrogen. Obtained cryo-samples were cut in 8 µm sections. Sections were stained by hematoxylin/eosin (HE) staining. Additionally, cryo-sections were submerged in 60% ethanol, 30% chloroform, 10% acetic acid solution for 30 min. Samples were thereafter counterstained by hematoxylin (5 s), followed by 800 mg/L Fast Green FCF staining (Merck; 10 min), 1% (v/v) acetic acid wash, and 1 g/L Safranin O (Merck) solution (10 min). Finally, stained sections were captured using Olympus IX70 converted microscope (Olympus, Tokyo, Japan).

2.10. Statistical Analysis

All data were processed, analyzed, and illustrated using GraphPad Prism (MacOS v9.0.1 (128), GraphPad Software, LLC, San Diego, CA, USA). Pictures and images were formatted using Adobe Illustrator (Adobe, San Diego, CA, USA). All values are presented as mean ±standard deviation. Statistical differences were determined via one-way ANOVA, followed by Dunnett's multiple or Tukey's comparison test. A p-value < 0.05 was considered statistically significant. Standard comparisons (as indicated by a *) involves the comparison of MSC samples cultured in identical conditions with the respective growth factor combination and the negative control of MSC cultured without growth factors.

3. Results

3.1. Alcian Blue Mediated Screening for Proteoglycan Production Rate

Human MSC were seeded in triplicate and subjected to culture with different growth factor combinations as a screening assay to identify the most promising combinations. (Figure 1) Specifically, GDF5 or GDF6 were combined with one or two factors from the selection; TGF-β1, TGF-β2, TGF-β3, Wnt3a, Wnt5a, Wnt11, and Shh. Following 3 weeks of culture, samples were examined via alcian blue staining. All growth factors except single factor supplementation of Shh and Wnt5a resulted in enhanced proteoglycan production (Figure 1). All other factors and factor combinations enhanced alcian blue deposition rates, although the rates diverged largely. Specifically, the combinations of TGF-β2/Wnt3a, TGF-β1/Wnt3a, GDF6/TGF-β1/Wnt3a, TGF-β3/Wnt3a, GDF6/TGF-β2/Wnt3a, TGF-β3/Wnt5a, GDF6/Wnt3a, GDF5/TGF-β3, GDF5/Wnt3a, and TGF-β1/Wnt5a proved most potent, respectively. The most potent growth factor combinations, as well as the Colombier GDF5/TGF-β1 combination [47], were further analyzed.

3.2. Nucleus Pulposus Cell Marker Expression Analysis

Stimulated MSC were cultured for 2 weeks. The differentiated cells were subsequently harvested and processed for mRNA expression analysis. Expression of NP cell maker [38] CD24 (Figure 2A), T (Figure 2B), SOX9 (Figure 2C), and COL2A1 (Figure 2D) all increased in MSC stimulated with growth factors, and approached expression levels of naturally obtained NP cells. Particularly the combinations TGF-β2/Wnt3a (p = 0.0522) showed strong enhancement in COL2A1 expression; however, no combinations resulted in a significant increase of COL2A1 levels. SOX9 expression was significantly higher in all conditions and highest in MSC treated with TGF-β1/Wnt3a (p < 0.0001) and TGF-β1/Wnt3a/GDF6 (p < 0.0001) combinations. Specifically, the combination GDF5/TGF-β1, GDF5/TGF-β3, TGF-β1/Wnt5a, and TGF-β3/Wnt5a were found to be inferior compared to GDF6/TGF-β1/Wnt3a, GDF6/TGF-β2/Wnt3a, TGF-β1/Wnt3a, and TGF-β3/Wnt3a. CD24 expression was significantly higher in all (but TGF-β3/Wnt5a) conditions, specifically, GDF6/Wnt3a (p < 0.0001) treated MSC. However, GDF6/Wnt3a showed significantly higher expressions compared to GDF5/TGF-β1, GDF5/TGF-β3, TGF-β1/Wnt5a, and TGF-β3/Wnt5a. TGF-β2/Wnt3a (p = 0.0009) and GDF6/Wnt3a (p = 0.0066) showed highest enhancement of T expression. TGF-β2/Wnt3a also showed superior T expression compared to GDF6/TGF-β2/Wnt3a and GDF5/Wnt3. As a result, we continued work with the combinations TGF-β2/Wnt3a, TGF-β1/Wnt3a, GDF6/TGF-β1/Wnt3a, and GDF6/Wnt3a, as these overall showed the strongest potency of NP cell marker induction.

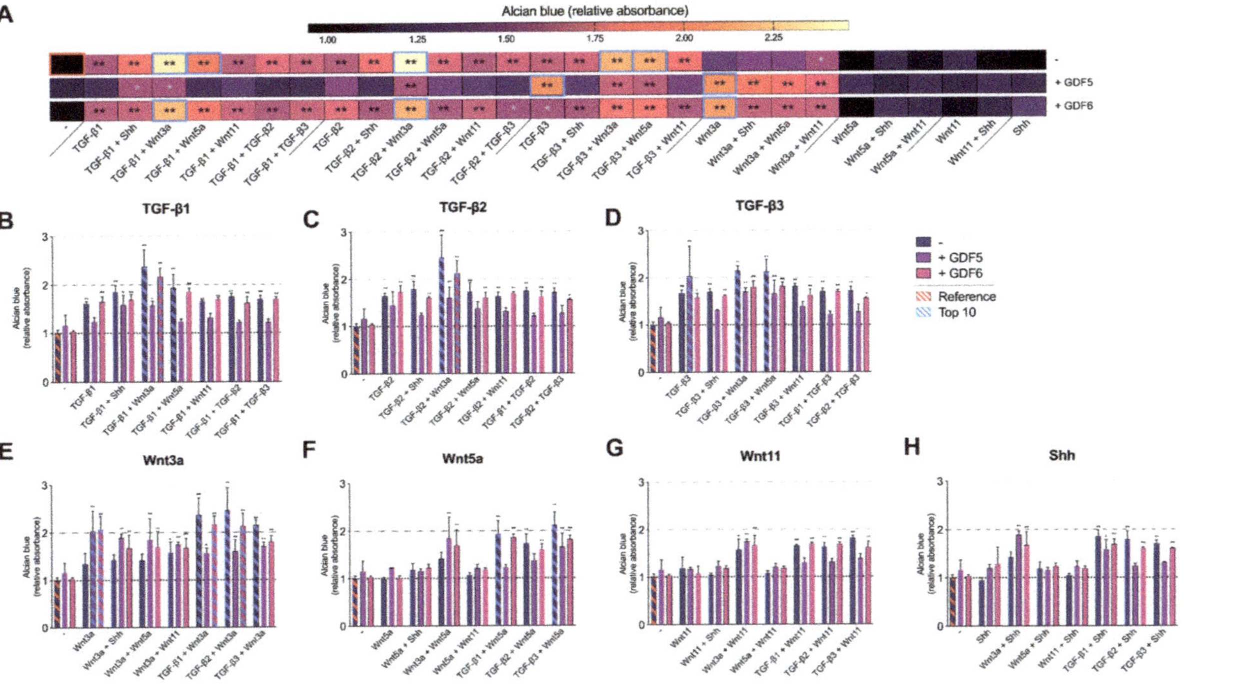

Figure 1. Representation of alcian blue screening assay for the assessment of different growth factor combinations to stimulate proteoglycan production in human MSC. Following 3 weeks of culture, alcian blue staining and subsequent absorbance measurement were used to determine the rate of proteoglycan production. (**A**) A large range of growth factor combinations without (top row) or with either GDF5 (middle row) or GDF6 (bottom row) to form a set of 1 to 3 growth factor media supplements were screened on their ability to stimulate MSC to produce proteoglycans. The top 10 combinations (highlighted with a blue border) were further assessed in subsequent assays. (**B–H**) Bar graph representation of heatmap data (to present variability in outcomes) of all combinations that contained (**B**) TGF-β1, (**C**) TGF-β2, (**D**) TGF-β3, (**E**) Wnt3a, (**F**) Wnt5a, (**G**) Wnt11, and (**H**) Shh. Additional supplementation of GDF5 (purple bars), GDF6 (pink bars), or without GDF (blue bars). The top 10 combinations are indicated by a blue pattern. Note that some combinations are repeatedly presented. All bars present mean values with error bars indicating standard deviations. Statistics were performed by comparing the average ($n = 3$) alcian blue absorbance values using one-way ANOVA followed by Dunnett's multiple comparison test where * indicates a $p < 0.05$, and ** a $p < 0.005$, comparing the values of the growth-factor-stimulated MSC to unstimulated MSC (-; indicated by red cell border or red pattern).

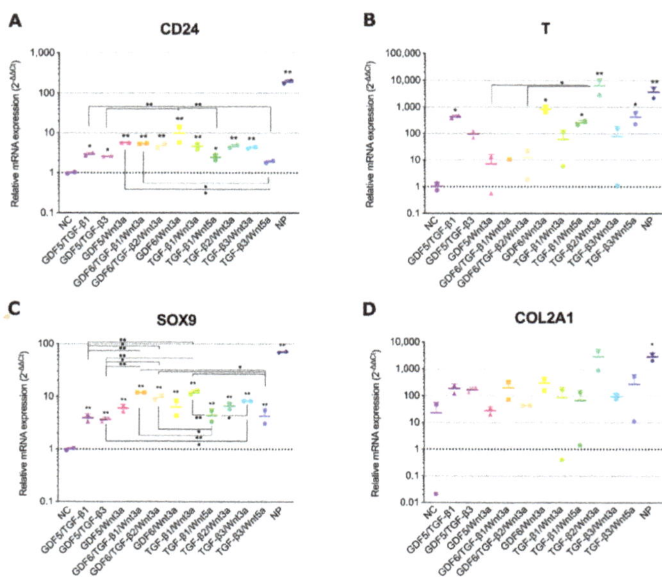

Figure 2. Gene expression assessment of NP cell markers in monolayer culture of MSC stimulated by different combinations of growth factors. (**A**) Relative expression levels of CD24 mRNA, (**B**) T (Brachyury) mRNA, (**C**) SOX9 mRNA, and (**D**) COL2A1 mRNA. Values are presented for individual donors (shapes) and as average values (bar), with error bars representing standard deviations. Statistical analysis was performed by one-way ANOVA followed by * Dunnett's multiple comparison test, which compared growth-factor-stimulated MSC to unstimulated MSC and • Tukey's multiple comparisons test comparing stimulated conditions. * or • indicates a $p < 0.05$, and ** or •• a $p < 0.005$. NC: negative control involving unstimulated MSC cultured in identical 2D culture conditions and media, but without growth factor supplementation, NP: natural human nucleus pulposus cells obtained from surgical waste tissue.

3.3. Chondrogenic Pellet Culture NP Cell Marker Expression

Next, the four top hitters in qPCR analysis, as well as the Colombier factors [47] of GDF5/TGF-β1, were applied for MSC differentiation in a standard chondrogenic 3D-pellet culture, to assess their ability to produce an NP-like chondrogenic matrix. Following 2 weeks of culture, pellets ($n = 3$) were harvested, RNA was isolated, and prepared for qPCR analysis. Expression levels were compared to MSC cultured without growth factor stimulation. CD24 expression (Figure 3A) revealed enhanced expression for all cases, with a significant increase for GDF5/TGF-β1 ($p < 0.0001$), TGF-β1/Wnt3a ($p = 0.0001$), and GDF6/TGF-β1/Wnt3a ($p < 0.0001$). Comparing the different combinations, we found GDF5/TGF-β1, GDF6/TGF-β1/Wnt3a, and TGF-β1/Wnt3a statistically outperforming GDF6/Wnt3a and TGF-β2/Wnt3a. SOX9 expression (Figure 3B) was similarly enhanced in all conditions, but most prominently for GDF5/TGF-β1 ($p = 0.0003$). COL2A1 (Figure 3C) showed similar results as the SOX9 assessment, with GDF5/TGF-β1 demonstrating the most prominent increase ($p < 0.0001$) and a significantly higher expression than all other combinations. Additionally, TGF-β1/Wnt3a and GDF6/TGF-β1/Wnt3a outperformed GDF6/Wnt3a and TGF-β2/Wnt3a. ACAN expression (Figure 3D) showed a trend of reduced expression in all conditions except GDF5/TGF-β1 and GDF6/TGF-β1/Wnt3a. Specifically, GDF5/TGF-β1 was the only combination that demonstrated an increase compared to the negative control, although not significant ($p = 0.0594$).

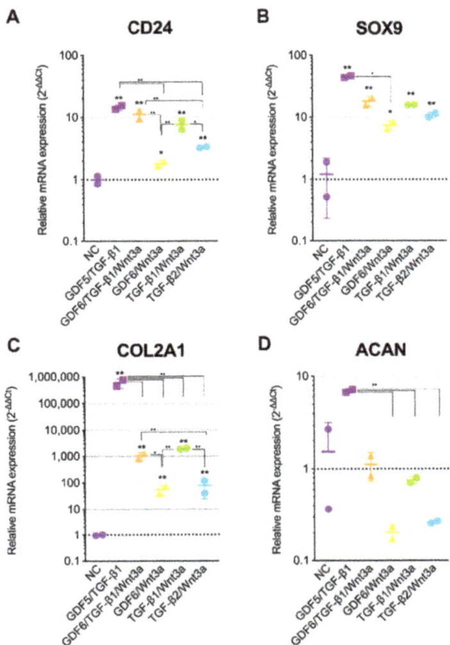

Figure 3. Gene expression assessment of NP cell markers within chondrogenic pellet culture containing human MSC stimulated by different combinations of growth factors. Relative expression levels of (**A**) CD24 mRNA, (**B**) SOX9 mRNA, (**C**) COL2A1 mRNA, and (**D**) ACAN mRNA. Values are presented for individual donors (shapes) and as average values (bar), with error bars representing standard deviations. Statistics were performed by one-way ANOVA followed by * Dunnett's multiple comparison test comparing growth-factor-stimulated to MSC cultured in identical conditions without growth factors (NC) and • Tukey's multiple comparisons test comparing stimulated conditions. * or • indicates a $p < 0.05$, and ** or •• a $p < 0.005$. NC: negative control involving unstimulated MSC cultured in identical pellet conditions and media, but without growth factor supplementation.

3.4. Chondrogenic Matrix Assessment Following MSC Differentiation

Following 3 weeks of culture, pellets were harvested either by papain digestion ($n = 3$) for DMMB and PicoGreen assay or processed for cryo-sectioning and subsequent histological analysis ($n = 3$). PicoGreen assay revealed a limited change in the number of cells compared to the negative control. (Figure 4B) Generally, most conditions showed a trend of a small increase in cell content, except for GDF6/TGF-β1/Wnt3a stimulated pellets. DMMB assay (Figure 4A) revealed an increase in sulfated glycosaminoglycans compared to the negative control, although the differences were not very large compared to the values obtained from NP cell-derived pellets. The GDF5/TGF-β1 combination showed the most potent in inducing glycosaminoglycan production, (5.974 (±1.850) times higher than the negative control; $p = 0.0496$). Finally, the ratio of glycosaminoglycans per DNA content (Figure 4C) showed the highest values for GDF5/TGF-β1 (4.926 ± 1.608) and GDF6/TGF-β1/Wnt3a (4.380 ± 2.461) stimulated pellets. However, only the GDF5/TGF-β1 combination involved a statistically significant increase with a $p = 0.0466$, while GDF6/TGF-β1/Wnt3a presented a $p = 0.1043$. No statistically significant differences were found between the growth factor conditions. Histological observations (Figure 4D) generally confirmed the DMMB findings. Safranin-O/Fast green stained sections revealed a matrix completely devoid of proteoglycans in the negative control (Figure 4D). Some safranin-O positive staining could be observed in the TGF-β2/Wnt3a, TGF-β1/Wnt3a, and GDF6/Wnt3a stimulated pellets (Figure 4D). This opposed to the pellet stimulated with

GDF5/TGF-β1 (Figures 4D,E and 5A), which showed regions of high-intensity safranin-O positive staining.

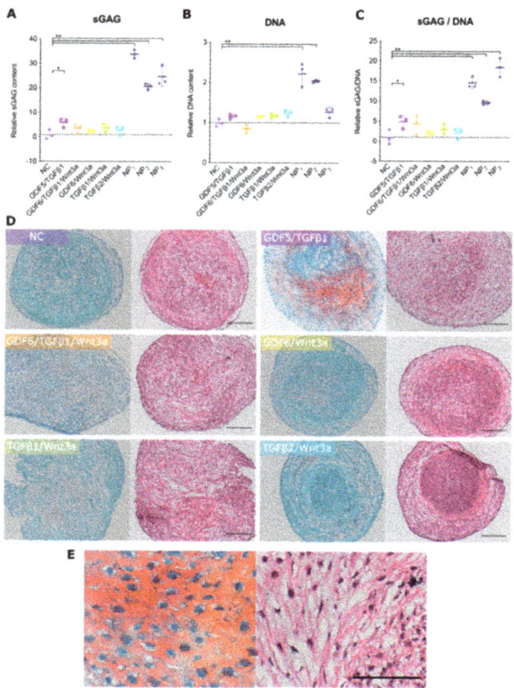

Figure 4. Assessment of growth factor combinations on extracellular matrix production of human MSC cultured in chondrogenic pellets. Via DMMB and Picogreen determined (**A**) glycosaminoglycan (sGAG), (**B**) DNA, and (**C**) the sGAG/DNA content was measured and calculated relative to non-stimulated MSC following 3 weeks of pellet culture ($n = 3$). (**D**) Histological examination of MSC-derived chondrogenic pellets stimulated with indicated growth factor combinations. Images represent pellets at 10× magnification stained with safranin-O/fastgreen (left) and hematoxylin/eosin (right). Scale bars represent 250 μm. (**E**) 40× magnification images of GDF5/TGF-β1. The scale bar equates to 125 μm. Statistics were performed by one-way ANOVA followed by Dunnett's multiple comparison test, where * indicates a $p < 0.05$ and ** a $p < 0.005$, relative to the NC (identically cultured, but without growth factors). No statistically significant differences were found between the different growth factor combinations. NC: negative control involves unstimulated MSC that were cultured in identical pellet conditions and media without growth factors, sGAG: sulfated glycosaminoglycans, NP: naturally obtained human nucleus pulposus cells from different donors.

3.5. Wnt3a Preconditioning to Enhance MSC Chondrogenesis

Human MSC expanded with or without 250ng mL^{-1} Wnt3a up to passage 1 and subsequently cultured for 3 weeks in GDF5/TGF-β1 containing differentiation media (Figure 5A,C) or commercial Chon.M. (Figure 5B,D). Histological observations did not reveal a clear benefit for MSC expanded in Wnt3a, as safranin-O staining was present at a similar intensity in pellets derived from both conditions.

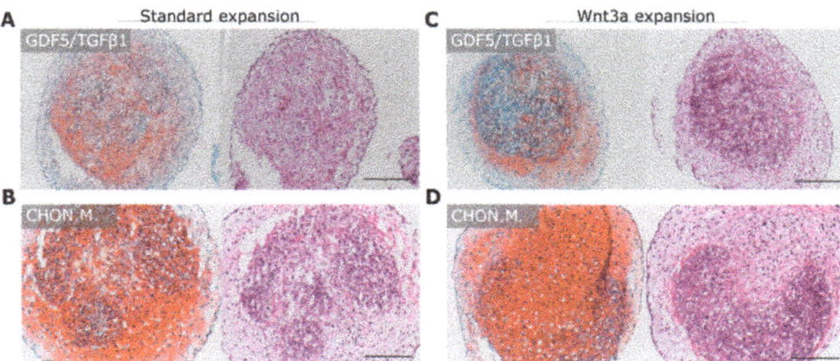

Figure 5. Assessment of Wnt3a pre-culturing for enhancement of chondrogenesis. Human MSC were expanded in standard expansion media without Wnt3a (**A**,**B**) or with 250 ng mL^{-1} Wnt3a (**C**,**D**). Subsequently obtained MSC were cultured in chondrogenic pellets and media supplemented with GDF5/TGF-β1 (**A**,**C**) or commercial predefined chondrogenic medium (**B**,**D**). Subsequent pellets were sectioned and stained by safranin-O/fast-green (left) or hematoxylin/eosin (right) staining. Scale bars represent 250 μm. Chon.M.: Mesenchymal stem Cell Chondrogenic Differentiation Medium.

4. Discussion

In this study, we aimed to identify growth factors or factor combinations that showed the highest potency in enhancing MSC to NP cell-like cell differentiation. Starting with monolayer cultures, a large array of factor combinations (86 different growth factors and factor combinations) were screened on their ability to stimulate proteoglycan production, as indicated by alcian blue staining. Unsurprisingly, almost all factor combinations showed the ability to at least enhance proteoglycan production compared to unstimulated MSC, as all factors were selected based on their general chondrogenic potential. Interestingly, of all the combinations identified as high hitters, a large fraction involved a combination with Wnt3a. A selection of the 10 most potent growth factor combinations, plus the combination of GDF5/TGF-β1, were further analyzed on their ability to induce NP cell marker expression. Here we showed the ability of particularly TGF-β2/Wnt3a, TGF-β1/Wnt3a, GDF6/TGF-β1/Wnt3a, and GDF6/Wnt3a to strongly enhance the expression of NP markers [38] CD24 and SOX9. A trend of enhancement in COL2A1 and T was observed for most conditions. Contrary to the impactful findings of Colombier et al. [47], our initial screening outcomes (in monolayer culture) did not indicate GDF5/TGF-β1 as a potent chondrogenic factor-combination.

Nevertheless, at this stage of the study, differentiation culture was still within the 2D monolayer set-up. MSC tend to change their characteristics relatively quickly in monolayer culture and lose their multipotency capacity [51–53]. Moreover, for induction of differentiation, the extracellular micro and macro-environment are critical as they can promote or support specific cellular behavior by providing mechanotransduction cues [54], anchor points, and chemical cues through the matrix proteins or by growth factor sequestering [55]. As such, we further validated the findings observed in monolayer cultures in a 3D environment, using a common pellet culture method [47]. Pellet cultures revealed an opposing trend, in which none of the Wnt3a combinations were able to significantly enhance glycosaminoglycan production rates. Although an increase was seen in the expression levels of some NP markers, i.e., CD24, SOX9, and COL2A1, increases in the ACAN expression level were not observed. This in contrast with the pellet cultures treated with GDF5/TGF-β1, which resulted in a significant increase in glycosaminoglycan production rates, and large proteoglycan positive regions within the pellet as indicated by safranin-O staining. Additionally, GDF5/TGF-β1 showed continuously most potent in enhancing NP cell marker expressions following the 3D pellet culture conditions. Although the relative glycosaminoglycan content of the other 4 growth factor combinations did not result in a significant increase, a trend of enhanced glycosaminoglycan production was observed both by the DMMB assay, as well as on histological observation. The negative control revealed no pink/red staining in the safranin-O stained section, the growth factor-stimulated pellets did present lightly pink areas, suggesting the presence of proteoglycans. Interestingly, the combination GDF6/TGF-β1/Wnt3a showed the second-highest rate of glycosaminoglycan production/DNA but did not strongly show safranin-O positive regions. The GDF6/TGF-β1/Wnt3a combination did present an overall lower DNA content, suggesting that the GDF6/TGF-β1/Wnt3a combination might have limited cellular proliferation or otherwise comprised cellular viability.

Although our initial screening failed to validate GDF5/TGF-β1 as a potent inducer of a chondrogenic and NP cell phenotype, subsequent assessment within a 3D environment did confirm the observations from Colombier et al. [47] that the synergy of GDF5/TGF-β1 was able to induce a desirable cell product. Even more, GDF5/TGF-β1 was the only combination able to result in fully chondrogenic pellets. The disagreement between 2D and 3D culture outcomes is not surprising. Cell behavior is strongly altered by such biomechanical cues [56]. Moreover, TGF-β1 is not a specific chondrogenic factor as it has similarly been shown to induce MSC differentiation to other cell lineages, e.g., cardiomyocytes, osteocytes, or adipocytes [57,58]. The ability to promote a chondrogenic NP cell-like phenotype through TGF-β1 is therefore likely context-dependent. This suggestion is supported by the original work of Colombier et al. [47], who showed the requirement of dexamethasone supplementation for a complete NP-like differentiation for MSC. That said, unlike the results of Colombier et al., our cells were unable to attain a final expression profile or matrix composition that matched those of naturally obtained human NP cells [47]. This might in part be due to the shorter culture periods that we employed. Colombier et al. maintained their pellet cultures for up to 4 weeks, whereas we cultured our pellets for 2 or 3 weeks dependent on the analysis performed. Additionally, while Colombier et al. applied adipogenic-derived MSC [47], our study employed bone marrow-derived MSC for the analysis. Distinct differentiation potential of differently sourced MSC is a well-reported phenomenon [46,59].

Wnt signaling is critical in general joint development, homeostasis, and has also been determined to have a role in disc degeneration [60–62]. Regardless, the role of Wnt3a in chondrogenesis is less clear [48]. Wnt3a has been shown able to enhance MSC proliferation rates and BMP2-mediated chondrogenesis, within 2D monolayer cultures [63]. Other studies have shown mixed results when applying Wnt3a on MSC for the enhancement of chondrogenic differentiation in 3D conditions [48]. Similarly, our initial screening confirmed these findings, where the highest-ranking growth factor combinations often included Wnt3a in monolayer culture. However, these enhancements were not observed in 3D culture. This could in part be ascribed to the potential role of Wnt3a and its maintenance of a more undifferentiated state, as suggested by Volleman et al. [48]. Where 2D culture could induce maturation or loss of potency with extended periods of culture, Wnt3a could potentially play a role in mitigating this loss of cell potency, thereby supporting the chondrogenic potential of additionally added growth factors. This beneficial role is not observed in 3D cultures. To assess this potential application of Wnt3a, we reapplied a pellet culture system, in which MSC were expanded with Wnt3a containing media to possibly maintain their potency, and subsequent differentiation in 3D pellet cultures with chondrogenic factors but without Wnt3a. The histological results were unable to confirm this theory, as the application of Wnt3a during expansion, did not appear to enhance the deposition rate of proteoglycan in the matrix.

Another aspect of our study is the direct comparison of growth factor combinations. For example, from our alcian blue assay (Figure 1), we observed that the addition of GDF5 appears to temper alcian blue rates compared to identical combinations without GDF5 or with GDF6 instead. Combinations that combined either Wnt11, Wnt5a, or Shh with GDF5 or GDF6 also did not present strong stimulation of proteoglycan production. Our screening assay also allows for further examination of specific combinations and single growth factor addition benefits or otherwise. For example, we found that the addition of either TGF-β1 or

TGF-β2 to GDF6/Wnt3a resulted in similar alcian blue production rates and subsequently did not appear to enhance NP marker expressions either. This suggested that TGF-β1 or TGF-β2 have no added value to the GDF6/Wnt3a combination. However, subsequent pellet cultures revealed that GDF6/Wnt3a with TGF-β1 resulted in significantly higher CD24 and COL2A1 expression levels, and showed a trend of enhanced ACAN and SOX9 expression levels. These observations further highlighting the context-dependent nature of cell culture and the effects of growth factors on the respective cells.

Notwithstanding, our results require some consideration before interpretation. One limitation of our study is the limited number of donor samples applied, particularly for the initial screening assay. Due to the high cost of applied growth factors and relatively long culture period, the initial screening of 86 different combinations was assessed on one single-sourced MSC. Moreover, the initial alcian blue screening was performed without correction for cell numbers or DNA content. Differences could in theory be caused by differences in cell content. However, we believe this is unlikely, as the long (3 weeks) culture period led all conditions to a confluent culture condition, and the differences at the time of assessment were likely minimal. Another study limitation to consider is that we did not assess outcomes of our monolayer and pellet cultures at different time points. Such particular combinations might have revealed more beneficial outcomes at shorter or longer stimulation periods. This requires further examination.

Overall, a need exists for novel regenerative strategies to treat intervertebral disc diseases. Currently, cell therapy [26] and growth factor injection [17] is gaining momentum, however, further validation and optimization of different regenerative products will likely be needed [64]. One potential approach for enhancing cellular products is priming the cells before transplantation [65], to promote an optimal anabolic state, such that the transplanted cells have the highest potential to contribute to the reorganization of the disc matrix or otherwise direct regional cells to do so [30]. Growth factors could be a potent tool in this regard. However, caution is required. Growth factors often function in context-dependent manners, and as such not every MSC population sourced from different donors might respond similarly. For clinical application and marketability, cell-product quality control is key, and this aspect should be taken into consideration [3,4,66]. Moreover, growth factors are generally quite expensive, particularly when applied over longer culture periods and for large-scale production. Thus, their application could hinder commercialization, by potentially increasing the price of cellular therapy production. Again, care should be taken that the benefits of growth factor stimulation outweigh the additional expense of the cellular product. From our screening of 86 different growth factor combinations for MSC to NP cell differentiation, we found that Wnt3a could potentially be an impactful factor when applied in 2D monolayer, although the mechanisms and requirements for these effects compel further exploration. Of all combinations analyzed, only the combination of GDF5 and TGF-β1 was able to result in fully chondrogenic pellets when applied in 3D cultures, and it might be the most potent combination for MSC to NP cell differentiation.

Author Contributions: D.S. and J.S.; conceptualization, T.N.E.V., K.M., J.S.; methodology, T.N.E.V., K.M., J.S.; formal analysis, D.S., M.S., M.W.; resources, J.S.; writing—original draft preparation, J.S., T.N.E.V., K.M., D.S., M.S., M.W.; writing—review and editing, J.S.; visualization, D.S., M.S., M.W.; supervision, D.S., M.S., M.W.; funding acquisition. All authors have read and agreed to the published version of the manuscript.

Funding: This research received no external funding.

Institutional Review Board Statement: The study was conducted according to the guidelines of the Declaration of Helsinki, and approved by the Institutional Review Board of the Tokai University School of Medicine (protocol code 06I-49 and 16R-051: 15 July 2016).

Informed Consent Statement: Informed consent was obtained from all subjects involved in the study.

Data Availability Statement: Data can be requested to the corresponding authors upon reasonable request.

Conflicts of Interest: The authors declare no conflict of interest.

Appendix A

Table A1. Tabular overview of primer sequences used for gene expression analysis.

Target	Primer Sequence (Forward)	Primer Sequence (Reverse)
GAPDH	5′-AATCAAGTGGGGCGATGCTG-′3	5′-GCAAATGAGCCCCAGCCTTC-′3
CD24	5′-GCACTGCTCCTACCCACGCAGATTT-′3	5′-GCCTTGGTGGTGGCATTAGTTGGAT-′3
Brachyury	5′-GGGTCCACAGCGCATGAT-′3	5′-TGATAAGCAGTCACCGCTATGAA-′3
COL2A1	5′-GGAAGAGTGGAGACTACTGGATTGAC-′3	5′-TCCATGTTGCAGAAAACCTTCA-′3
SOX9	5′-CAGACAGCCCCCTATCGACT-′3	5′-CGTTGACATCGAAGGTCTCG-′3
ACAN	5′-GTGCCTATCAGGACAAGGTCT-′3	5′-GATGCCTTTCACCACGACTTC-′3

References

1. Vos, T.; Allen, C.; Arora, M.; Barber, R.M.; Bhutta, Z.A.; Brown, A.; Carter, A.; Casey, D.C.; Charlson, F.J.; Chen, A.Z.; et al. Global, regional, and national incidence, prevalence, and years lived with disability for 310 diseases and injuries, 1990–2015: A systematic analysis for the Global Burden of Disease Study 2015. *Lancet* **2016**, *388*, 1545–1602. [CrossRef]
2. Schol, J.; Sakai, D. Cell therapy for intervertebral disc herniation and degenerative disc disease: Clinical trials. *Int. Orthop.* **2019**, *43*, 1011–1025. [CrossRef] [PubMed]
3. Smith, L.J.; Silverman, L.; Sakai, D.; Le Maitre, C.L.; Mauck, R.L.; Malhotra, N.R.; Lotz, J.C.; Buckley, C.T. Advancing cell therapies for intervertebral disc regeneration from the lab to the clinic: Recommendations of the ORS spine section. *JOR Spine* **2018**, *1*, e1036. [CrossRef] [PubMed]
4. Thorpe, A.A.; Bach, F.C.; Tryfonidou, M.A.; Le Maitre, C.L.; Mwale, F.; Diwan, A.D.; Ito, K. Leaping the hurdles in developing regenerative treatments for the intervertebral disc from preclinical to clinical. *JOR Spine* **2018**, *1*, e1027. [CrossRef]
5. Oichi, T.; Taniguchi, Y.; Oshima, Y.; Tanaka, S.; Saito, T. Pathomechanism of intervertebral disc degeneration. *JOR Spine* **2020**, *3*, e1076. [CrossRef]
6. Vergroesen, P.P.; Kingma, I.; Emanuel, K.S.; Hoogendoorn, R.J.; Welting, T.J.; van Royen, B.J.; van Dieën, J.H.; Smit, T.H. Mechanics and biology in intervertebral disc degeneration: A vicious circle. *Osteoarthr. Cartil.* **2015**, *23*, 1057–1070. [CrossRef]
7. Zeldin, L.; Mosley, G.E.; Laudier, D.; Gallate, Z.S.; Gansau, J.; Hoy, R.C.; Poeran, J.; Iatridis, J.C. Spatial mapping of collagen content and structure in human intervertebral disk degeneration. *JOR Spine* **2020**, *3*, e1129. [CrossRef]
8. Fournier, D.E.; Kiser, P.K.; Shoemaker, J.K.; Battie, M.C.; Seguin, C.A. Vascularization of the human intervertebral disc: A scoping review. *JOR Spine* **2020**, *3*, e1123. [CrossRef]
9. Vincent, K.; Dona, C.P.G.; Albert, T.J.; Dahia, C.L. Age-related molecular changes in the lumbar dorsal root ganglia of mice: Signs of sensitization, and inflammatory response. *JOR Spine* **2020**, *3*, e1124. [CrossRef]
10. Ma, J.; Stefanoska, D.; Stone, L.S.; Hildebrand, M.; van Donkelaar, C.C.; Zou, X.; Basoli, V.; Grad, S.; Alini, M.; Peroglio, M. Hypoxic stress enhances extension and branching of dorsal root ganglion neuronal outgrowth. *JOR Spine* **2020**, *3*, e1090. [CrossRef]
11. Risbud, M.V.; Shapiro, I.M. Role of cytokines in intervertebral disc degeneration: Pain and disc content. *Nat. Rev. Rheumatol.* **2014**, *10*, 44–56. [CrossRef]
12. Zhang, S.; Hu, B.; Liu, W.; Wang, P.; Lv, X.; Chen, S.; Shao, Z. The role of structure and function changes of sensory nervous system in intervertebral disc-related low back pain. *Osteoarthr. Cartil.* **2021**, *29*, 17–27. [CrossRef]
13. Fritzell, P.; Hagg, O.; Nordwall, A.; Swedish Lumbar Spine Study Group. Complications in lumbar fusion surgery for chronic low back pain: Comparison of three surgical techniques used in a prospective randomized study. A report from the Swedish Lumbar Spine Study Group. *Eur. Spine J.* **2003**, *12*, 178–189. [CrossRef]
14. Zhong, Z.M.; Deviren, V.; Tay, B.; Burch, S.; Berven, S.H. Adjacent segment disease after instrumented fusion for adult lumbar spondylolisthesis: Incidence and risk factors. *Clin. Neurol. Neurosurg.* **2017**, *156*, 29–34. [CrossRef]
15. North American Spine Society. *Diagnosis and Treatment of Low Back Pain*; NASS: Burr Ridge, IL, USA, 2020.
16. Krupkova, O.; Cambria, E.; Besse, L.; Besse, A.; Bowles, R.; Wuertz-Kozak, K. The potential of CRISPR/Cas9 genome editing for the study and treatment of intervertebral disc pathologies. *JOR Spine* **2018**, *1*, e1003. [CrossRef]
17. Hodgkinson, T.; Shen, B.; Diwan, A.; Hoyland, J.A.; Richardson, S.M. Therapeutic potential of growth differentiation factors in the treatment of degenerative disc diseases. *JOR Spine* **2019**, *2*, e1045. [CrossRef]
18. Bowles, R.D.; Setton, L.A. Biomaterials for intervertebral disc regeneration and repair. *Biomaterials* **2017**, *129*, 54–67. [CrossRef]
19. Harmon, M.D.; Ramos, D.M.; Nithyadevi, D.; Bordett, R.; Rudraiah, S.; Nukavarapu, S.P.; Moss, I.L.; Kumbar, S.G. Growing a backbone—Functional biomaterials and structures for intervertebral disc (IVD) repair and regeneration: Challenges, innovations, and future directions. *Biomater. Sci.* **2020**, *8*, 1216–1239. [CrossRef]

20. Gullbrand, S.E.; Smith, L.J.; Smith, H.E.; Mauck, R.L. Promise, progress, and problems in whole disc tissue engineering. *JOR Spine* **2018**, *1*, e1015. [CrossRef]
21. Buckley, C.T.; Hoyland, J.A.; Fujii, K.; Pandit, A.; Iatridis, J.C.; Grad, S. Critical aspects and challenges for intervertebral disc repair and regeneration-Harnessing advances in tissue engineering. *JOR Spine* **2018**, *1*, e1029. [CrossRef]
22. Sakai, D.; Andersson, G.B. Stem cell therapy for intervertebral disc regeneration: Obstacles and solutions. *Nat. Rev. Rheumatol.* **2015**, *11*, 243–256. [CrossRef] [PubMed]
23. Hiraishi, S.; Schol, J.; Sakai, D.; Nukaga, T.; Erickson, I.; Silverman, L.; Foley, K.; Watanabe, M. Discogenic cell transplantation directly from a cryopreserved state in an induced intervertebral disc degeneration canine model. *JOR Spine* **2018**, *1*, e1013. [CrossRef] [PubMed]
24. Nukaga, T.; Sakai, D.; Schol, J.; Sato, M.; Watanabe, M. Annulus fibrosus cell sheets limit disc degeneration in a rat annulus fibrosus injury model. *JOR Spine* **2019**, *2*, e1050. [CrossRef] [PubMed]
25. Omlor, G.W.; Lorenz, S.; Nerlich, A.G.; Guehring, T.; Richter, W. Disc cell therapy with bone-marrow-derived autologous mesenchymal stromal cells in a large porcine disc degeneration model. *Eur. Spine J.* **2018**, *27*, 2639–2649. [CrossRef]
26. Sakai, D.; Schol, J. Cell therapy for intervertebral disc repair: Clinical perspective. *J. Orthop. Transl.* **2017**, *9*, 8–18. [CrossRef]
27. Musial-Wysocka, A.; Kot, M.; Majka, M. The Pros and Cons of Mesenchymal Stem Cell-Based Therapies. *Cell Transpl.* **2019**, *28*, 801–812. [CrossRef]
28. Nukaga, T.; Sakai, D.; Schol, J.; Suyama, K.; Nakai, T.; Hiyama, A.; Watanabe, M. Minimal Sustainability of Dedifferentiation by ROCK Inhibitor on Rat Nucleus Pulposus Cells In Vitro. *Spine Surg. Relat. Res.* **2019**, *3*, 385–391. [CrossRef]
29. Sakai, D.; Schol, J.; Bach, F.C.; Tekari, A.; Sagawa, N.; Nakamura, Y.; Chan, S.C.W.; Nakai, T.; Creemers, L.B.; Frauchiger, D.A.; et al. Successful fishing for nucleus pulposus progenitor cells of the intervertebral disc across species. *JOR Spine* **2018**, *1*, e1018. [CrossRef]
30. Loibl, M.; Wuertz-Kozak, K.; Vadala, G.; Lang, S.; Fairbank, J.; Urban, J.P. Controversies in regenerative medicine: Should intervertebral disc degeneration be treated with mesenchymal stem cells? *JOR Spine* **2019**, *2*, e1043. [CrossRef]
31. Vickers, L.; Thorpe, A.A.; Snuggs, J.; Sammon, C.; Le Maitre, C.L. Mesenchymal stem cell therapies for intervertebral disc degeneration: Consideration of the degenerate niche. *JOR Spine* **2019**, *2*, e1055. [CrossRef]
32. Amirdelfan, K.; Bae, H.; McJunkin, T.; DePalma, M.; Kim, K.; Beckworth, W.J.; Ghiselli, G.; Bainbridge, J.S.; Dryer, R.; Deer, T.R.; et al. Allogeneic mesenchymal precursor cells treatment for chronic low back pain associated with degenerative disc disease: A prospective randomized, placebo-controlled 36-month study of safety and efficacy. *Spine J* **2021**, *21*, 212–230. [CrossRef]
33. Ni, L.; Liu, X.; Sochacki, K.R.; Ebraheim, M.; Fahrenkopf, M.; Shi, Q.; Liu, J.; Yang, H. Effects of hypoxia on differentiation from human placenta-derived mesenchymal stem cells to nucleus pulposus-like cells. *Spine J* **2014**, *14*, 2451–2458. [CrossRef]
34. Bertolo, A.; Mehr, M.; Aebli, N.; Baur, M.; Ferguson, S.J.; Stoyanov, J.V. Influence of different commercial scaffolds on the in vitro differentiation of human mesenchymal stem cells to nucleus pulposus-like cells. *Eur. Spine J.* **2012**, *21* (Suppl. S6), S826–S838. [CrossRef]
35. Frauchiger, D.A.; Heeb, S.R.; May, R.D.; Woltje, M.; Benneker, L.M.; Gantenbein, B. Differentiation of MSC and annulus fibrosus cells on genetically engineered silk fleece-membrane-composites enriched for GDF-6 or TGF-beta3. *J. Orthop. Res.* **2018**, *36*, 1324–1333. [CrossRef]
36. Zhang, Y.; Zhang, Z.; Chen, P.; Ma, C.Y.; Li, C.; Au, T.Y.K.; Tam, V.; Peng, Y.; Wu, R.; Cheung, K.M.C.; et al. Directed Differentiation of Notochord-like and Nucleus Pulposus-like Cells Using Human Pluripotent Stem Cells. *Cell Rep.* **2020**, *30*, 2791–2806. [CrossRef]
37. Bucher, C.; Gazdhar, A.; Benneker, L.M.; Geiser, T.; Gantenbein-Ritter, B. Nonviral Gene Delivery of Growth and Differentiation Factor 5 to Human Mesenchymal Stem Cells Injected into a 3D Bovine Intervertebral Disc Organ Culture System. *Stem Cells Int.* **2013**, *2013*, 326828. [CrossRef]
38. Risbud, M.V.; Schoepflin, Z.R.; Mwale, F.; Kandel, R.A.; Grad, S.; Iatridis, J.C.; Sakai, D.; Hoyland, J.A. Defining the phenotype of young healthy nucleus pulposus cells: Recommendations of the Spine Research Interest Group at the 2014 annual ORS meeting. *J. Orthop. Res.* **2015**, *33*, 283–293. [CrossRef]
39. Narcisi, R.; Cleary, M.A.; Brama, P.A.; Hoogduijn, M.J.; Tüysüz, N.; ten Berge, D.; van Osch, G.J. Long-term expansion, enhanced chondrogenic potential, and suppression of endochondral ossification of adult human MSCs via WNT signaling modulation. *Stem Cell Rep.* **2015**, *4*, 459–472. [CrossRef]
40. Dickinson, S.C.; Sutton, C.A.; Brady, K.; Salerno, A.; Katopodi, T.; Williams, R.L.; West, C.C.; Evseenko, D.; Wu, L.; Pang, S.; et al. The Wnt5a Receptor, Receptor Tyrosine Kinase-Like Orphan Receptor 2, Is a Predictive Cell Surface Marker of Human Mesenchymal Stem Cells with an Enhanced Capacity for Chondrogenic Differentiation. *Stem Cells* **2017**, *35*, 2280–2291. [CrossRef]
41. Narcisi, R.; Quarto, R.; Ulivi, V.; Muraglia, A.; Molfetta, L.; Giannoni, P. TGF beta-1 administration during ex vivo expansion of human articular chondrocytes in a serum-free medium redirects the cell phenotype toward hypertrophy. *J. Cell Physiol.* **2012**, *227*, 3282–3290. [CrossRef]
42. Pei, M.; Li, J.; Zhang, Y.; Liu, G.; Wei, L.; Zhang, Y. Expansion on a matrix deposited by nonchondrogenic urine stem cells strengthens the chondrogenic capacity of repeated-passage bone marrow stromal cells. *Cell Tissue Res.* **2014**, *356*, 391–403. [CrossRef] [PubMed]
43. Stoyanov, J.V.; Gantenbein-Ritter, B.; Bertolo, A.; Aebli, N.; Baur, M.; Alini, M.; Grad, S. Role of hypoxia and growth and differentiation factor-5 on differentiation of human mesenchymal stem cells towards intervertebral nucleus pulposus-like cells. *Eur. Cell Mater.* **2011**, *21*, 533–547. [CrossRef] [PubMed]

44. Clarke, L.E.; McConnell, J.C.; Sherratt, M.J.; Derby, B.; Richardson, S.M.; Hoyland, J.A. Growth differentiation factor 6 and transforming growth factor-beta differentially mediate mesenchymal stem cell differentiation, composition, and micromechanical properties of nucleus pulposus constructs. *Arthritis Res. Ther.* **2014**, *16*, R67. [CrossRef] [PubMed]
45. Hodgkinson, T.; Stening, J.Z.; White, L.J.; Shakesheff, K.M.; Hoyland, J.A.; Richardson, S.M. Microparticles for controlled growth differentiation factor 6 delivery to direct adipose stem cell-based nucleus pulposus regeneration. *J. Tissue Eng. Regen. Med.* **2019**, *13*, 1406–1417. [CrossRef]
46. Hodgkinson, T.; Wignall, F.; Hoyland, J.A.; Richardson, S.M. High BMPR2 expression leads to enhanced SMAD1/5/8 signalling and GDF6 responsiveness in human adipose-derived stem cells: Implications for stem cell therapies for intervertebral disc degeneration. *J. Tissue Eng.* **2020**, *11*, 2041731420919334. [CrossRef]
47. Colombier, P.; Clouet, J.; Boyer, C.; Ruel, M.; Bonin, G.; Lesoeur, J.; Moreau, A.; Fellah, B.H.; Weiss, P.; Lescaudron, L.; et al. TGF-beta1 and GDF5 Act Synergistically to Drive the Differentiation of Human Adipose Stromal Cells toward Nucleus Pulposus-like Cells. *Stem Cells* **2016**, *34*, 653–667. [CrossRef]
48. Volleman, T.N.E.; Schol, J.; Morita, K.; Sakai, D.; Watanabe, M. Wnt3a and wnt5a as Potential Chondrogenic Stimulators for Nucleus Pulposus Cell Induction: A Comprehensive Review. *Neurospine* **2020**, *17*, 19–35. [CrossRef]
49. Mochida, J.; Sakai, D.; Nakamura, Y.; Watanabe, T.; Yamamoto, Y.; Kato, S. Intervertebral disc repair with activated nucleus pulposus cell transplantation: A three-year, prospective clinical study of its safety. *Eur. Cell Mater.* **2015**, *29*, 202–212. [CrossRef]
50. Gibson, J.D.; O'Sullivan, M.B.; Alaee, F.; Paglia, D.N.; Yoshida, R.; Guzzo, R.M.; Drissi, H. Regeneration of Articular Cartilage by Human ESC-Derived Mesenchymal Progenitors Treated Sequentially with BMP-2 and Wnt5a. *Stem Cells Transl. Med.* **2017**, *6*, 40–50. [CrossRef]
51. Neuhuber, B.; Swanger, S.A.; Howard, L.; Mackay, A.; Fischer, I. Effects of plating density and culture time on bone marrow stromal cell characteristics. *Exp. Hematol.* **2008**, *36*, 1176–1185. [CrossRef]
52. Li, X.Y.; Ding, J.; Zheng, Z.H.; Li, X.Y.; Wu, Z.B.; Zhu, P. Long-term culture in vitro impairs the immunosuppressive activity of mesenchymal stem cells on T cells. *Mol. Med. Rep.* **2012**, *6*, 1183–1189. [CrossRef]
53. Kretlow, J.D.; Jin, Y.Q.; Liu, W.; Zhang, W.J.; Hong, T.H.; Zhou, G.; Baggett, L.S.; Mikos, A.G.; Cao, Y. Donor age and cell passage affects differentiation potential of murine bone marrow-derived stem cells. *BMC Cell Biol.* **2008**, *9*, 60. [CrossRef]
54. Fearing, B.V.; Hernandez, P.A.; Setton, L.A.; Chahine, N.O. Mechanotransduction and cell biomechanics of the intervertebral disc. *JOR Spine* **2018**, *1*. [CrossRef]
55. Schultz, G.S.; Wysocki, A. Interactions between extracellular matrix and growth factors in wound healing. *Wound Repair Regen.* **2009**, *17*, 153–162. [CrossRef]
56. Ravalli, S.; Szychlinska, M.A.; Lauretta, G.; Musumeci, G. New Insights on Mechanical Stimulation of Mesenchymal Stem Cells for Cartilage Regeneration. *Appl. Sci.* **2020**, *10*, 2927. [CrossRef]
57. Kurpinski, K.; Lam, H.; Chu, J.; Wang, A.; Kim, A.; Tsay, E.; Agrawal, S.; Schaffer, D.V.; Li, S. Transforming growth factor-beta and notch signaling mediate stem cell differentiation into smooth muscle cells. *Stem Cells* **2010**, *28*, 734–742. [CrossRef]
58. Elsafadi, M.; Manikandan, M.; Almalki, S.; Mobarak, M.; Atteya, M.; Iqbal, Z.; Hashmi, J.A.; Shaheen, S.; Alajez, N.; Alfayez, M.; et al. TGFbeta1-Induced Differentiation of Human Bone Marrow-Derived MSCs Is Mediated by Changes to the Actin Cytoskeleton. *Stem Cells Int.* **2018**, *2018*, 6913594. [CrossRef]
59. Mohamed-Ahmed, S.; Fristad, I.; Lie, S.A.; Suliman, S.; Mustafa, K.; Vindenes, H.; Idris, S.B. Adipose-derived and bone marrow mesenchymal stem cells: A donor-matched comparison. *Stem Cell Res. Ther.* **2018**, *9*, 168. [CrossRef]
60. Hiyama, A.; Sakai, D.; Risbud, M.V.; Tanaka, M.; Arai, F.; Abe, K.; Mochida, J. Enhancement of intervertebral disc cell senescence by WNT/beta-catenin signaling-induced matrix metalloproteinase expression. *Arthritis Rheumatol.* **2010**, *62*, 3036–3047. [CrossRef]
61. Hiyama, A.; Sakai, D.; Tanaka, M.; Arai, F.; Nakajima, D.; Abe, K.; Mochida, J. The relationship between the Wnt/beta-catenin and TGF-beta/BMP signals in the intervertebral disc cell. *J. Cell Physiol.* **2011**, *226*, 1139–1148. [CrossRef]
62. Usami, Y.; Gunawardena, A.T.; Iwamoto, M.; Enomoto-Iwamoto, M. Wnt signaling in cartilage development and diseases: Lessons from animal studies. *Lab. Investig.* **2016**, *96*, 186–196. [CrossRef]
63. Fischer, L.; Boland, G.; Tuan, R.S. Wnt-3A enhances bone morphogenetic protein-2-mediated chondrogenesis of murine C3H10T1/2 mesenchymal cells. *J. Biol. Chem.* **2002**, *277*, 30870–30878. [CrossRef] [PubMed]
64. Croft, A.S.; Guerrero, J.; Oswald, K.A.C.; Häckel, S.; Albers, C.E.; Gantenbein, B. Effect of different cryopreservation media on human nucleus pulposus cells' viability and trilineage potential. *JOR Spine* **2021**, e1140. [CrossRef] [PubMed]
65. Elabd, C.; Centeno, C.J.; Schultz, J.R.; Lutz, G.; Ichim, T.; Silva, F.J. Intra-discal injection of autologous, hypoxic cultured bone marrow-derived mesenchymal stem cells in five patients with chronic lower back pain: A long-term safety and feasibility study. *J. Transl. Med.* **2016**, *14*, 253. [CrossRef] [PubMed]
66. Silverman, L.I.; Flanagan, F.; Rodriguez-Granrose, D.; Simpson, K.; Saxon, L.H.; Foley, K.T. Identifying and Managing Sources of Variability in Cell Therapy Manufacturing and Clinical Trials. *Regen. Eng. Transl. Med.* **2019**, *5*, 354–361. [CrossRef]

Article

Optimization of Spheroid Colony Culture and Cryopreservation of Nucleus Pulposus Cells for the Development of Intervertebral Disc Regenerative Therapeutics

Kosuke Sako [1], Daisuke Sakai [1,*], Yoshihiko Nakamura [2], Erika Matsushita [2], Jordy Schol [2], Takayuki Warita [2], Natsumi Horikita [2], Masato Sato [1] and Masahiko Watanabe [1,*]

1 Department of Orthopaedic Surgery, Tokai University School of Medicine, 143 Shimokasuya, Isehara, Kanagawa 259-1193, Japan; k.sako0626@tokai.ac.jp (K.S.); sato-m@is.icc.u-tokai.ac.jp (M.S.)

2 Research Center for Regenerative Medicine, Tokai University School of Medicine, 143 Shimokasuya, Isehara, Kanagawa 259-1193, Japan; kahiko@is.icc.u-tokai.ac.jp (Y.N.); silsilsilring@gmail.com (E.M.); bb.jordy@gmail.com (J.S.); takayuki.warita@tunzpharma.co.jp (T.W.); natsumi.horikita@tunzpharma.co.jp (N.H.)

* Correspondence: daisakai@is.icc.u-tokai.ac.jp (D.S.); masahiko@is.icc.u-tokai.ac.jp (M.W.)

Citation: Sako, K.; Sakai, D.; Nakamura, Y.; Matsushita, E.; Schol, J.; Warita, T.; Horikita, N.; Sato, M.; Watanabe, M. Optimization of Spheroid Colony Culture and Cryopreservation of Nucleus Pulposus Cells for the Development of Intervertebral Disc Regenerative Therapeutics. *Appl. Sci.* **2021**, *11*, 3309. https://doi.org/10.3390/app11083309

Academic Editor: Andrea Ballini

Received: 28 February 2021
Accepted: 1 April 2021
Published: 7 April 2021

Publisher's Note: MDPI stays neutral with regard to jurisdictional claims in published maps and institutional affiliations.

Abstract: After the discovery of functionally superior Tie2-positive nucleus pulposus (NP) progenitor cells, new methods were needed to enable mass culture and cryopreservation to maintain these cells in an undifferentiated state with high cell yield. We used six types of EZSPHERE® dishes, which support spheroid-forming colony culture, and examined NP cell spheroid-formation ability, number, proliferation, and mRNA expression of *ACAN*, *COL1A2*, *COL2A1*, and *ANGPT1*. Six different types of cryopreservation solutions were examined for potential use in clinical cryopreservation by comparing the effects of exposure time during cryopreservation on cell viability, Tie2-positivity, and cell proliferation rates. The spheroid formation rate was 45.1% and the cell proliferation rate was 7.75 times using EZSPHERE® dishes. The mRNA levels for *COL2A1* and *ANGPT1* were also high. In cryopreservation, CryoStor10 (CS10) produced ≥90% cell viability and a high proliferation rate after thawing. CS10 had a high Tie2-positive rate of 12.6% after culturing for 5 days after thawing. These results suggest that EZSPHERE enabled colony formation in cell culture without the use of hydrogel products and that CS10 is the best cryopreservation medium for retaining the NP progenitor cell phenotype and viability. Together, these data provide useful information of NP cell-based therapeutics to the clinic.

Keywords: intervertebral disc; nucleus pulposus; nucleus pulposus progenitor cells; cryopreservation; Tie2; spheroid colony; EZSPHERE; CryoStor 10

1. Introduction

Low back pain is the primary cause of disability worldwide and often leads to poor quality of life for patients [1,2]. Low back pain is thought to relate to intervertebral disc (IVD) degeneration, although the mechanisms have not been clearly identified [3]. The IVD is composed of three distinct tissues: the nucleus pulposus (NP), annulus fibrosus (AF), and cartilage endplate. The NP comprises the gelatinous tissue located in the center of the IVD. The AF comprises layered collagen structures that surround the NP [4]. Previous reports have indicated that IVD homeostasis is determined by the NP [5,6] and that IVD degeneration is thought to be caused by dysregulation of extracellular matrix (ECM) homeostasis within the NP tissue [7,8].

Disc degeneration is associated with a decline in NP cell numbers and potency, and renewal by stem or progenitor sources appears to be limited [9]. Interestingly, in 2012, Sakai et al. discovered a specific NP progenitor cell population species [9]. NP progenitor cells were shown to have the ability for self-renewal and multipotency, including adipogenic, osteogenic, and neurogenic differentiation potential [9–12]. NP progenitor cells were reported to differentiate into two types of colony-forming units upon culture in methylcellulose semisolid medium: spheroid colony-forming units (CFU-s) with high ECM production and fibroblastic colony-forming units (CFU-f) with low functionality [9,13]. The propensity of NP progenitor cells for CFU-s was linked to the expression of the cell surface marker angiopoietin-1 receptor (Tie2) [9]. Tie2-expressing NP progenitor cells have since been identified in human, canine, murine, and bovine IVD samples [12–14]. Interestingly, the presence of Tie2-expressing cells within the IVD declines sharply with aging [9] and is similarly linked with progression of disc degeneration [9]. Thus, Tie2 progenitor cells provide a promising therapeutic target in the treatment of IVD degeneration.

Cell therapy is a rapidly evolving field and is thought to be safe and effective for the treatment of low back pain associated with degenerating IVDs [15] Cell therapy has been suggested to be effective through repopulation of the IVD cells or reactivation of endemic cells through paracrine signaling, thereby switching the disc environment from catabolic to anabolic [15]. Our group has conducted one of the first IVD cell therapy clinical trials, which involved the transplantation of autologous NP cells that were reactivated using coculture with mesenchymal stem cells (MSCs) [16]. The trial revealed that NP cells could be safely transplanted and that cell transplantation appears to be effective in preventing disc degeneration [16].

Although these results are promising, NP materials are often derived from aged and diseased patients, which can compromise their regenerative potential [17–19]. We believe that the potency of NP cell therapy could be increased by using a cell population with high Tie2-positive rates. However, Tie2-positive cells remain difficult to maintain in culture because they rapidly lose their Tie2 expression and consequently their proliferative capacity and potency [13]. In addition, mass production of these cells is limited by the high viscosity of the methylcellulose medium and unsuitability for industrialization.

A variety of studies have investigated ways to improve the culture conditions [10,14,20] or isolation methods [21] to increase the number of Tie2-expressing cells. For example, Tekari et al. reported beneficial effects of fibroblast growth factor 2 (FGF2) and hypoxic culture conditions [14]. Recent work by Guerrero et al. [10] showed the benefits of three-dimensional (3D) pellet culture environments for maintaining the multipotent capacity and retaining Tie2 expression in Tie2-positive NP progenitor cells. These findings provide evidence for the potential of culture augmentation to increase the retention of NP progenitor cells. However, an optimized culture method that would be scalable, marketable, and clinically applicable remains to be determined.

Another consideration that could hinder the translation of Tie2-positive NP progenitor cells to the market is their preservation. Tanaka et al. has shown that cell viability, differentiation ability, and ECM production do not change significantly from before to after NP cell cryopreservation [22]. This finding suggests that activated NP cells could be used in a cryopreserved state whenever needed according to the patient's condition, thus increasing their potential as a commercial, mass-produced, off-the-shelf (OTS) regenerative therapeutic product. In a canine study, an OTS discogenic NP cell product was shown to be as effective in regenerating induced disc degeneration compared with freshly cultured cells [23]. These products are currently undergoing clinical trial evaluations (NCT03347708, NCT03955315).

There is a large difference between the actual clinical application and the results in laboratory-based reports. In addition, the time required before and after freezing is rarely considered [24]. The cytotoxicity of animal serum and dimethyl sulfoxide (DMSO) contained in the cryopreservation solutions may also present safety and regulatory problems during clinical applications [13,25]. Furthermore, during the product management and transportation after mass culture, the process from cell isolation to storage becomes more complicated and time consuming.

We wanted to examine the potential for optimization of the practical aspects of NP cell production in terms of cell culture, storage, and logistic handling. Given the needs of future commercialization, we aimed to identify optimal cryopreservation solution(s) that could limit the functional deterioration and maintain the high cell viability of Tie2-positive NP cells. We examined the potential for using specialized nonadherent EZSPHERE® dishes, which have been used with induced pluripotent stem cells and cancer cells but not with NP cells. This culture dish supports 3D cell cultures without the need for hydrogel encapsulation to enable mass production of spheroid-forming NP cells.

2. Materials and Methods

2.1. Human NP Cell Isolation

Human NP tissues were collected from 10 patients undergoing surgery for lumbar disc herniation at Tokai University Hospital and related facilities (Table 1). All patients provided their informed consent for the collection and use of surgical waste for research purposes. All research procedures described in this study were approved by the Institutional Review Board for Clinical Research, Tokai University (application number 16R-051: 15 July 2016).

Table 1. Details of clinical samples. Age, sex, total weight of IVD tissue collected at the time of surgery, and weight of NP tissue after selection are listed for each sample. LDH: lumber disc hernia, IVD: intervertebral disc, NP: nucleus pulposus.

	Code	Age	Sex	Disease	IVD Tissue Amount(g)	NP Tissue Amount(g)
1	T19	19	M	LDH	8.05	7.59
2	A29	29	F	LDH	3.76	2.57
3	K30	30	F	LDH	4.26	2.5
4	E32	32	F	LDH	4.75	4.75
5	A16	16	F	LDH	3.58	1.9
6	K21	21	F	LDH	7.52	4.79
7	A24	24	M	LDH	8.04	4.69
8	T23	23	M	LDH	5.06	4.6
9	T18	18	M	LDH	3.38	3
10	A18	18	M	LDH	5.88	3.73

NP cells were isolated as described previously [13]. Briefly, the collected NP tissue was washed with excess saline and the tissue wet weight was measured. In a 100 mm dish, the tissue was cut into 3–5 mm-sized pieces with sterilized scissors and scalpels. The shredded NP tissue was carefully transferred to a 50 mm conical tube and centrifuged at 1200 rpm for 5 min at 4 °C. After centrifugation, the supernatant was removed, and the tissue was resuspended in 10 mL of 10% (*v*/*v*) fetal bovine serum (FBS), minimal essential medium α (MEMα, Fujifilm Wako Pure Chemical Corporation, Osaka, Japan), and 10 mL of TrypLE Express (Thermo Fisher Scientific, Tokyo, Japan). The suspension was incubated with gentle swirling at 37 °C for 1 h. The tissue dissolution state was confirmed and the sample was centrifuged at 1800 rpm for 5 min. Next, the tissue was resuspended in 15 mL of 10% FBS–MEMα and 5 mL of 0.25 mg/mL collagenase P (Roche, Basel, Switzerland), and incubated for 2 h at 37 °C. After digestion, the tissue was again centrifuged and resuspended in 20 mL of 10% FBS–MEMα, the cell suspension was filtered through a 40 μm cell strainer (Corning, Corning, NY, USA), and the cells were subjected to experimental cryopreservation or culture conditions as described below.

2.2. NP Cell Culture Conditions

Primary human NP cells (*n* = 4, Table 1: T19, A29, K30, E32) were suspended at 10,000 cells/mL in 10% FBS–MEMα. Cells were seeded in 2 mL onto different types of EZSPHERE® plates (spheroid-formation culture vessel: IWAKI, Tokyo, Japan) and cultured at 37 °C in 5% CO_2 and 5% O_2 for 14 days. EZSPHERE® plates have a uniform well structure on the surface of the dish that allow cells to form different shapes of germ layers or spheroids depending on the size and type of the well (Figure 1). Here we used the following EZSPHERE® dish types: 4000-900 (caliber: 500 μm, depth: 100 μm, number of wells: 2300); 4000-901 (caliber: 200 μm, depth: 100 μm, number of wells: 9200);

4000-902 (diameter: 500 μm, depth: 200 μm, number of wells: 2300); 4000-903 (diameter: 800 μm, depth: 400 μm, number of wells: 600), and 4000-904 (diameter: 800 μm, depth: 300 μm), number of wells: 200); type 4000-905 (caliber: 1400 μm, depth: 600 μm, number of wells: 2300). After 2 weeks of culture, the established colonies were collected carefully by pipetting. For analysis of marker expression, standard monolayer culture was performed by seeding NP cells at a density of 4.5×10^5 cells/cm^2 into 100 mm plates and culturing them under identical conditions for 7 days. These cells were used as the control.

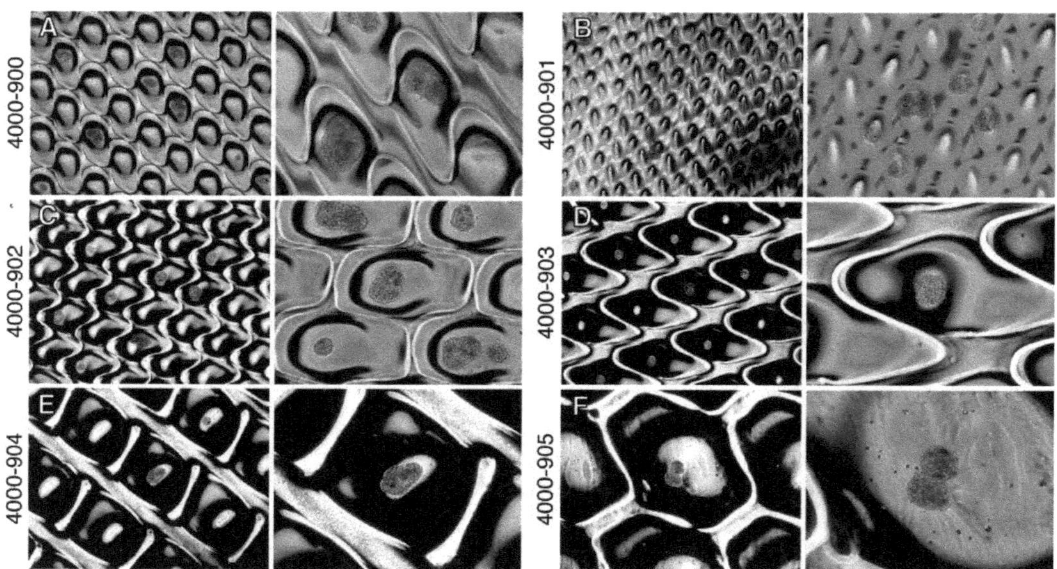

Figure 1. Cells cultured giving rise to spheroid forming units within EZSPHERE® dish (**A**) type 4000-900, (**B**) type 4000-901, (**C**) type 4000-902, (**D**) type 4000-903, (**E**) type 4000-904, and (**F**) type 4000-905 at 4× (**left**) and 10× (**right**) magnification. Images visualized by inverted phase contrast microscope. 20,000 human nucleus pulposus cells were seeded per dish and cultured for 14 days at 5% CO_2, 5% O_2, and 37 °C.

NP cells (n = 1, donor E32) were also cultured in EZSPHERE® dish type 4000-903 with normal medium or medium supplemented with 100 ng/mL epidermal growth factor (EGF; R&D Systems, Minneapolis, MN, USA), 10 ng/mL FGF2 (PeproTech, Cranbury, NJ, USA), 100 ng/mL platelet-derived growth factor (PDGF; R&D Systems), or a combination of all three factors. The colonies obtained were similarly harvested and evaluated.

2.3. *Cryopreservation of NP Cells*

After confirming they were free of contamination and the state of the cells, primary NP cells were treated with 5 mL of TrypLE Express for 3 min and then collected into a 15 mL conical tube. Samples were centrifuged at 500× g for 5 min at 4 °C, resuspended in 5 mL of buffered saline, and centrifuged again at 500× g for 5 min at 4 °C. The cells were counted using the trypan blue exclusion method to determine their cell number and viability.

For the first experiment, the cell suspension (donor T19) was centrifuged once more to prepare a 1 mL suspension of CELLBANKER 1 (CB1), STEM-CELLBANKER GMP grade (SCB), or STEM-CELLBANKER DMSO Free GMP grade (SCBD-Free) (all from Zenoaq Resource, Fukushima, Japan) (Table 2), each containing 1×10^6 cells in their respective cryopreservation solution to be dispensed into a cryotube. The samples were kept at room temperature or on ice for 0.5, 1, 2, 3, 4, and 5 h, and cell viability was directly analyzed either or analyzed following cryostorage at −80 °C for 1 day. Cryostorage was prepared with

BICELL containers (Nihon Freezer, Tokyo, Japan) in same way as previously reported [22] to control the cooling rate.

Table 2. Details of cryopreservation solutions. Whether include the animal serum, clears GMP grade, and concentration of DMSO are listed for each cryopreservation. CB-1: CELLBANKER 1, SCB: STEM-CELLBANKER GMP grade, SCBD-free: STEM-CELLBANKER DMSO Free GMP grade, CS10: CryoStor® CS10, and PF: Pro Freeze™-CDM.

Cryopreservation	Animal Serum	Concetration of DMSO	GMP Grade
CB1	Included	10%	Not Cleared
SCB	Not Included	10%	Cleared
SCBD-free	Not Included	0%	Cleared
CS10	Not Included	10%	Cleared
CP-1™	Not Included	5%	Not Cleared
PF	Not Included	15%	Not Cleared

Next, the effects of cryopreservation solutions on NP cell potency and surface marker expression were examined. Based on the previous step, a 3 h incubation time on ice was selected. NP cells (n = 3, T19, A24, and T23) were aliquoted at 1×10^6 cells in 1 mL of CB1, SCB, or SCBD-Free. The cells were then frozen at –80 °C for 1 day, thawed, and cultured in monolayer culture for 5 days. The cells were retrieved using TrypLE Express and analyzed by flow cytometry.

Tie2 expression and cell viability of NP cells (n = 1, donor, T19) were examined using six different cryopreservation compositions: CB1, SCB, SCBD-Free, CryoStor® CS10 (CS10) (Stemcell Technologies, Vancouver, BC, Canada), CP-1™ (Kyokuto Pharmaceutical Industry, Tokyo, Japan), and Pro Freeze™-CDM (PF) (Lonza Bioscience, Basel, Switzerland). The final tubes were incubated on ice, frozen at −80 °C, and cultured before being analyzed as previously described.

Finally, 1×10^6 NP cells (n = 3, T19, A18, and T18) were kept in 1 mL of CB1 or CS10 on ice or at room temperature for 0, 2, 3, or 5 h. Cells were either analyzed directly or cryopreserved at –80 °C for 1 day.

2.4. Thawing Protocol of NP Cells

Cells cryopreserved in cryotubes were slowly warmed in a bath with 37 °C water for about one minute. The tube was removed from the pre-heated water and kept on room temperature until complete thawing was observed. When thawed, the solution was carefully transferred to a falcon tube, and dropwise media containing 10% FBS was added to slowly resuspend the cells and dilute the cryo-storage media. The samples were then centrifuged at 1200 rpm for 5 min at 4 °C to collect the cells and applied for the respective experiments.

2.5. Colony Formation and Cell Proliferation Rates

After the cells were cultured for 14 days in EZSPHERE® plates, an inverted phase-contrast microscope was used to count the number of colonies that had formed per 100 wells. To calculate the number of cells, the entire colonies were collected, centrifuged, and treated with 1% Liberase (Roche) for 30 min at 37 °C. The cells were centrifuged again and resuspended, and the number of cells derived from the colonies was counted.

2.6. Cell Viability

Cell viability was measured using the standard trypan blue exclusion method and an inverted phase-contrast microscope; the percentages of living and dead cells were counted.

2.7. Real-Time Polymerase Chain Reaction (PCR) Analysis

Cultured human NP cells (n = 3, E32, A16 m, and K21) were homogenized using ISO-GEN II (Nippon Gene, Tokyo, Japan), and RNA was prepared using an RNAqueous Micro RNA isolation kit (Applied Biosystems, Thermo Fisher Scientific, MA, USA) following the manufacturer's instructions. High-Capacity cDNA Reverse Transcription Kits (Applied Biosystems 4387406) were used to reverse transcribe mRNA and to synthesize cDNA. Based on TaqMan gene expression assays (Applied Biosystems), gene expression levels were detected using TaqMan Universal PCR Master Mix and the following primers: TaqMan Gene Expression (human): ACAN (Hs00153936; Applied Biosystems), COL1A2 (Hs01028956; Applied Biosystems), COL2A1 (Hs00264051; Applied Biosystems), and ANGPT1 (Hs00919202; Applied Biosystems). PCR was performed using an amplification machine (7500; RealTime PCR Systems) with stage 1 (95 °C for 20 s × 1 cycle) and stage 2 (95 °C for 3 s, 60 °C for 30 s × 50 cycles). C_T values were calculated relative to the C_T value for the housekeeping gene GAPDH (4333764F; Applied Biosystems).

2.8. Flow Cytometry and Fluorescence-Activated Cell Sorting

NP cells were analyzed using a FACSCalibur flow cytometer (BD Biosciences, Franklin Lakes, NJ, USA). Only living cells were targeted by using a living (propidium iodide-negative) gate. The cell sorting methods and monoclonal antibodies used for analysis were performed as described previously [13].

2.9. Statistical Analysis

All data were collected and processed using Microsoft Excel (Microsoft, Redmond, WA, USA). Statistical analysis was performed using IBM SPSS Statistics software (version 26; IBM Corp., Armonk, NY, USA). Significant differences were identified using the Mann–Whitney U test and Kruskal–Wallis test. All values are presented as mean (±standard deviation). p values of <0.05 were considered to be significant.

3. Results

3.1. Spheroid Colony-Formation and Cell Proliferation Rates in EZSPHERE Dishes

The ability of EZSPHERE dishes to support NP cell colony formation was evaluated by determining the colony formation and cell proliferation rates (Figure 1). Interestingly, the colonies formed in each plate were similar in size (Figure 1). The average colony formation rates were 9.1% (±3.0%) for type 4000-901, 15.6% (±9.6%) for type 4000-900, 24.0% (±1.0%) for type 4000-902, 45.1% (±12.1%) for type 4000-903, 17.5% (±7.0%) for type 4000-904, and 15.1% (±5.1%) for type 4000-905. Dish 4000-903 significantly outperformed almost all other dish types (Figure 2A).

Similarly, the cell proliferation rates were highest using the 400-903 plate, although the only significant difference was compared with 4000-905 (p = 0.029) (Figure 2B). Dish type 4000-903 seemed to be the most suitable type for NP cell CFU culture. Next, growth factor supplementation showed strongly increased proliferation rates by up to 45 times when stimulated with FGF2 or 62.5 times when stimulated with all three factors combined (Figure 2C).

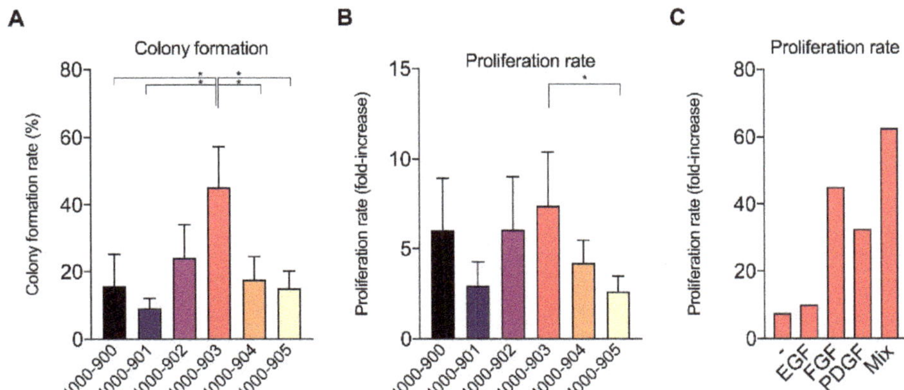

Figure 2. (**A**) The rate of spheroid colonies that formed and (**B**) fold increase of cells on different EZSPHERE® dish types derived from 20,000 human nucleus pulposus cells ($n = 4$) following 14 days of culture. (**C**) The fold-increase in cell numbers following culture of nucleus pulposus cells ($n = 1$) on EZSPHERE® dish type 4000-903, without (-) or with 100 ng/mL epidermal growth factor (EGF), 10 ng/mL fibroblast growth factor (FGF), 100 ng/mL platelet derived growth factor (PDGF), and a combination of all three growth factors (Mix). Data is presented as mean values with error bars presenting standard deviation. * $p < 0.05$,

3.2. NP Cell Marker Expression in EZSPHERE Spheroid Colonies

Real-time PCR was performed after culturing cells in each of the six types of EZSPHERE dishes. The expression levels of ANGPT1 and COL2A1 were higher in 3D compared with 2D culture (Figure 3A,D). The COL2A1 expression was particularly prominent, which supports previous reports of high ECM production performance in colony-forming cells [9]. Augmentation of EZSPHERE culture conditions by growth factor supplementation revealed strong reduction in the expression levels of ACAN, COL1A2, COL2A1, and ANGPT1. (Figure 3E–H). Notably, the standard deviation for type 4000-901 was observably high for COL2A1 and ANGPT1 outcomes, predominantly due to high expression found for the A29 donor. (Figure 3A,D) The expression levels for the FGF2- and PDGF-supplemented conditions were similar to those of the monolayer-cultured NP cells.

3.3. Cell Viability after Addition of Cryopreservation Solution

To examine the effects of different cryopreservation solutions on NP cells during mass cell batch production, we tested extended exposure of NP cells to three types of cryopreservation solution (CB1, SCB, and SCBD-Free). At room temperature, cells exposed to SCB and SCBD-Free showed a rapid decline in cell viability, with >50% cell loss within 1 h (Figure 4A). In the same process on ice, a high cell viability of ≥80% was obtained in all solutions for up to 2 h, but the number of viable cells decreased with longer incubations in SCB and SCBD-Free (Figure 4C). Samples left at room temperature for ≥1 h followed by cryopreservation showed complete loss of viability (Figure 4B). Samples kept on ice before cryopreservation had better viability, although cell viability was 67% in SCB and 57.8% in SCBD-Free conditions after 2 h and decreased further with time (Figure 4D). Cells exposed to CB1 had a survival rate of ≥90% even after being left for 5 h on ice, although the cell viability decreased with time at room temperature.

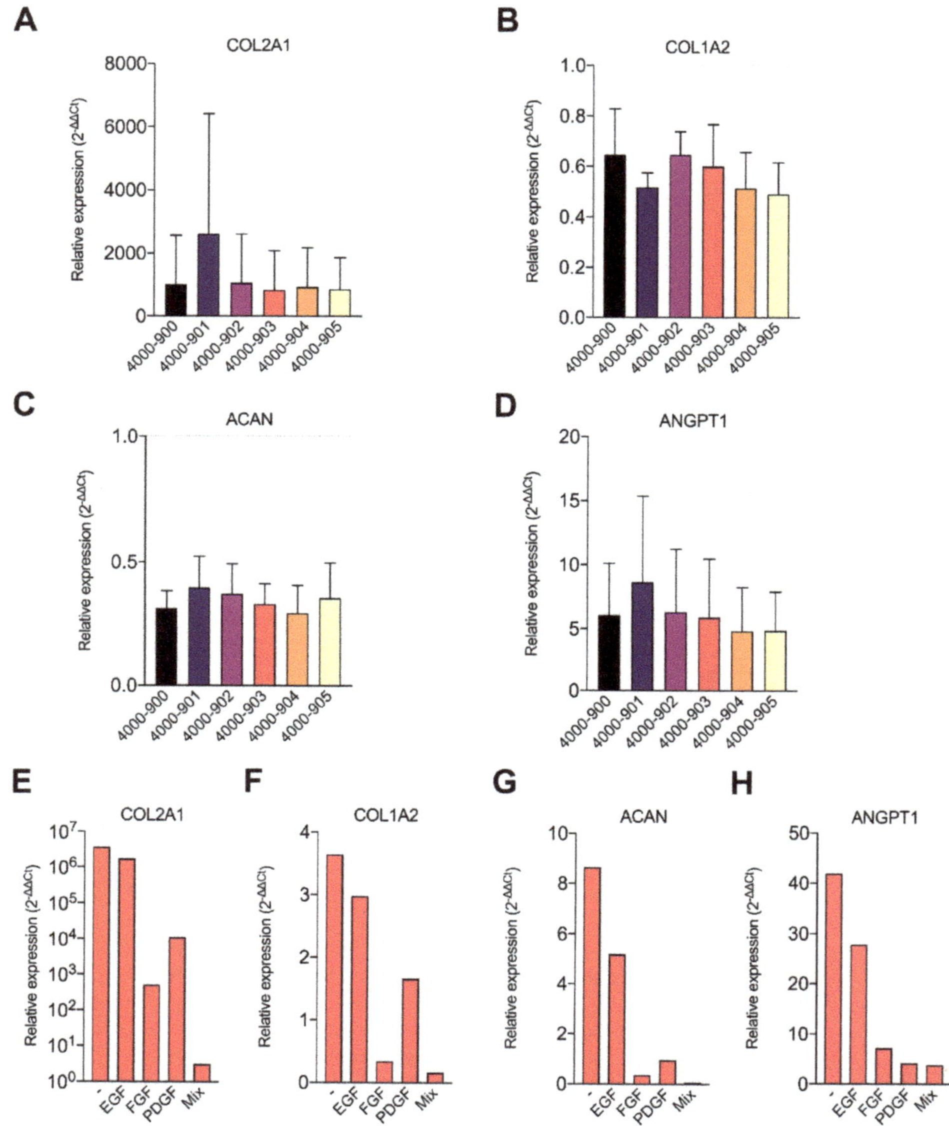

Figure 3. Relative gene expression levels of (**A**) COL2A1, (**B**) COL1A2, (**C**) ACAN, and (**D**) ANGPT1 for nucleus pulposus cell spheroid colonies cultured on different EZSPHERE® dish types. (n = 4) Relative gene expression of (**E**) COL2A1, (**F**) COL1A2, (**G**) ACAN, and (**H**) ANGPT1 of nucleus pulposus cell cultured on EZSPHERE® dish 4000-903 (n = 1) without (-) or with 100 ng/mL epidermal growth factor (EGF), 10 ng/mL fibroblast growth factor (FGF), 100 ng/mL platelet derived growth factor (PDGF), or a combination of all three growth factors (Mix). Data is calculated relative to monolayer cultured nucleus pulposus cells. Values derived using $2^{-\Delta\Delta Ct}$ method and presented as mean values with error bars presenting standard deviation.

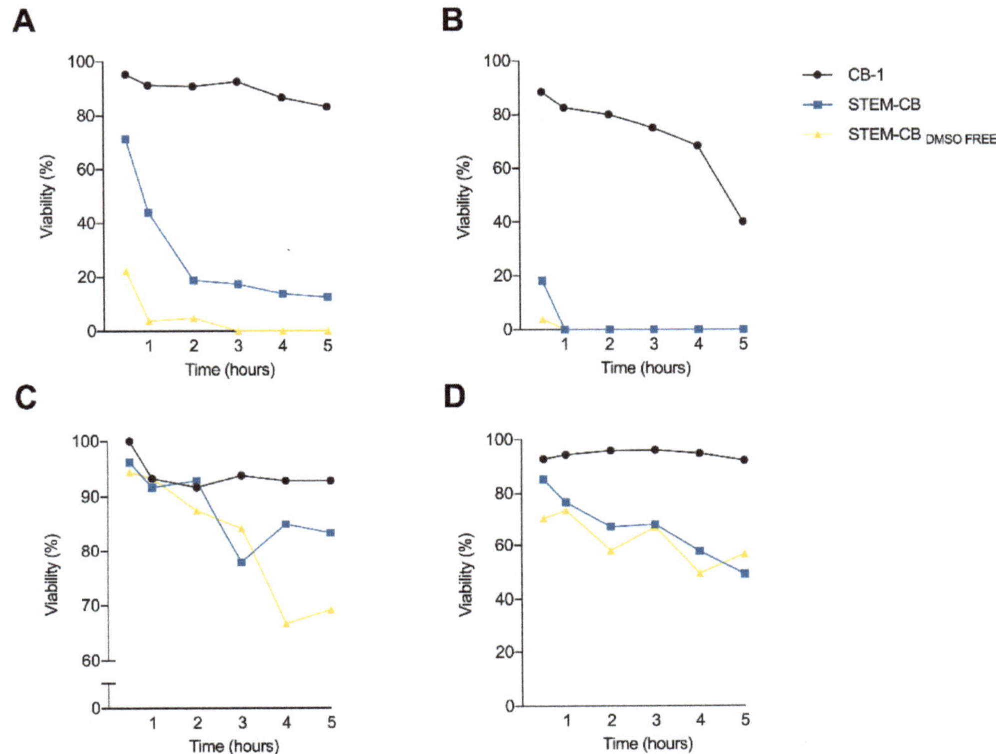

Figure 4. Cells (n = 1) were suspended in CellBanker-1 (CB-1), STEM-CELLBANKER (STEM-CB), or STEM-CELLBANKER without DMSO (STEM-CB$_{\text{DMSO FREE}}$) and kept for up to 5 h on (**A**,**B**) room temperature or kept (**C**,**D**) on ice. Cell viability tested at multiple time points either before or after 1 day of cryopreservation.

3.4. Cell Potential after Cryopreservation

Next, NP cells were exposed to cryostorage solutions (CB1, SCB, or SCBD-Free) for 3 h on ice before freezing. The cells were thawed and subsequently cultured. The cells frozen in CB1 had an average cell viability of 95.7% (Figure 4A), but the other two cryopreservation solutions resulted in considerable cell loss (Figure 5A). The average growth rate of cells frozen in CB1 was 5.9 times (p = 0.53), which was lower than the growth rate of cells frozen in SCB (8.93 times; Figure 5B). The Tie2- and CD24-positive rates in flow cytometry analysis showed a trend toward reduced Tie2- and CD24-expressing fractions for NP cells kept in CB 1 compared with SCBD-Free, although this difference was not significant (Figure 5C,D).

Additional investigation with a wider range of cryopreservation solutions (CB1, SCB, SCBD-Free, CS10, CP-1, and PF) revealed viability ranging from 63.6% (PC) to 84.6% (CS10) (Figure 6A). Interestingly, the Tie2-positivity rates of NP cells was 12.6% in cells in CS10, suggesting a higher Tie2 retention rate in cells in CS10 than in CB1 (Figure 6B).

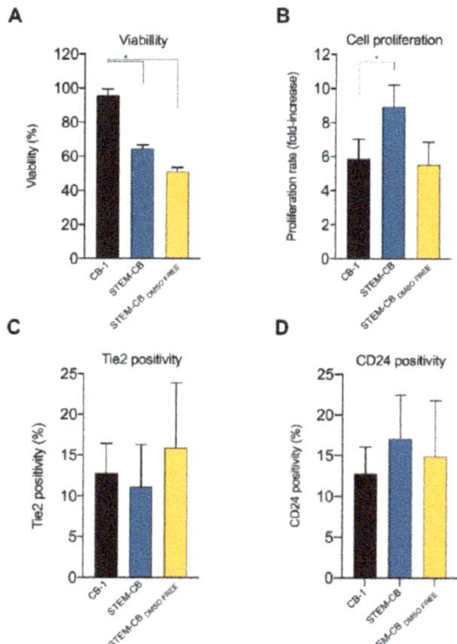

Figure 5. Cells ($n = 3$) were suspended in CellBanker-1 (CB-1), STEM-CELLBANKER (STEM-CB), or STEM-CELLBANKER without DMSO (STEM-$CB_{DMSO\ FREE}$) and kept for on ice for 3 h and frozen for 1 day. Subsequently (**A**) cell viability was assessed. Following cryostorage and thawing cells were seeded cultured for 5 days. The (**B**) fold-increase in cell numbers was analyzed, as well as the fraction of (**C**) Tie2 expressing and (**D**) CD24 expressing cells as analyzed by flow cytometry. Data is presented as mean values with error bars presenting standard deviation. * $p < 0.05$.

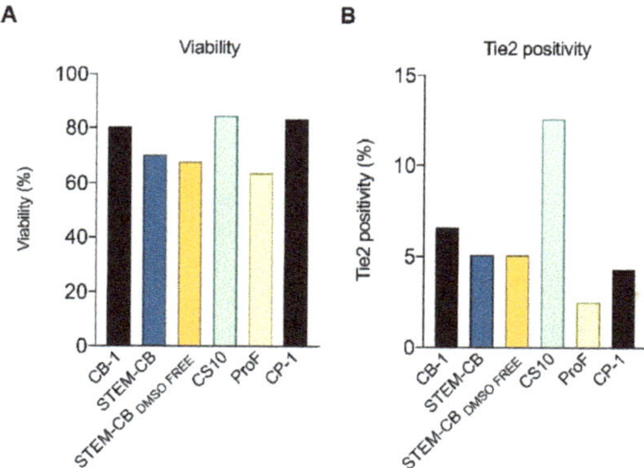

Figure 6. Comparison of 6 types of cryopreservation solutions, namely CELLBANKER 1 (CB-1), STEM-CELLBANKER GMP grade (STEM-CB), STEM-CELLBANKER DMSO Free GMP grade (STEM-$CB_{DMSO\ FREE}$), CryoStor® CS10 (CS10), and Pro Freeze™-CDM (ProF). Nucleus pulposus cells suspended in the cryopreservation solutions and kept on ice for 3 h before cryostoring them for 1 day. The cells were recovered and cultured for 5 days. Subsequently, the (**A**) viability rate and the (**B**) fraction of Tie2 positive cells were determined. Data is presented as mean values with error bars presenting standard deviation.

Finally, CS10 and CB1 were used to resuspended primary NP cells and the effects on cell viability in cells with time left at room temperature or on ice were examined. The average cell viability was 95.2% (±1.8%) for cells in CS10, and this value was maintained even after 5 h on ice (Figure 7A) and at 89.9% (6.5%) in cells stored at room temperature

(Figure 7B). CB1 produced similar results (Figure 6B). The average cell viability after cryopreservation and thawing also remained high: 88.4% (±4.6%) in cells on ice in CS10 or 87.7% (±6.0%) in CB1 (Figure 7C). However, when kept at room temperature, the survival rate after cryopreservation and thawing appeared to decrease to 67.5% (±29.9%) after 3 h in CS10 compared with 82.1% (±6.5%) in CB1 (Figure 7D).

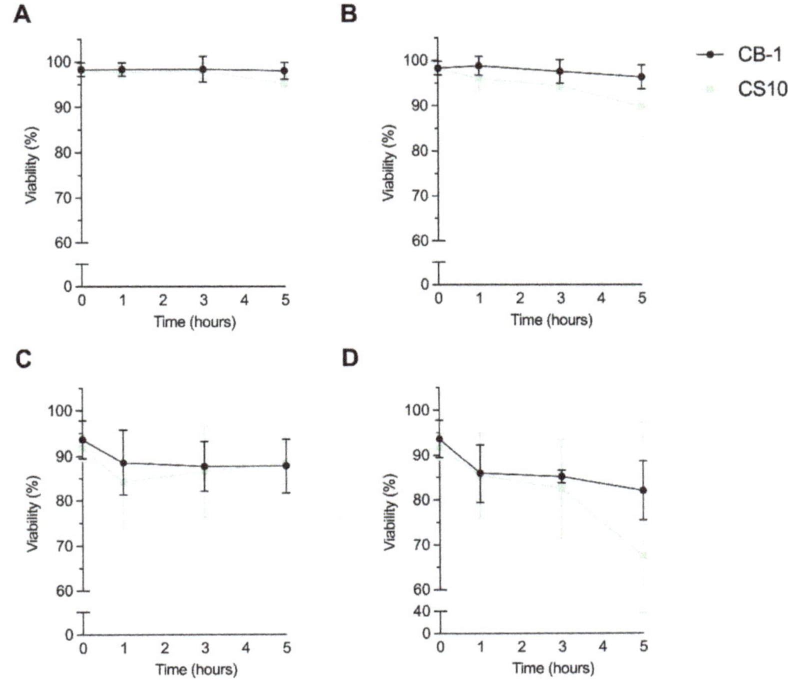

Figure 7. Cells (*n* = 3) were suspended in CellBanker-1 (CB-1) and CryoStor-10 (CS10) and kept for up to 5 h on (**B**,**D**) room temperature or (**A**,**C**) on ice. Cell viability tested at multiple time points either before or after 1 day of cryopreservation.

4. Discussion

In this study, we aimed to tackle some practical issues that potentially limit the market and clinical translation of NP progenitor cells as a regenerative cell transplantation product. First, we examined the ability of EZSPHERE cultures plates to allow large-scale colony-forming culture of NP progenitor cells. Second, we examined the effects of cryopreservation solution on NP cell viability and potency under real-world conditions.

Spheroid formation rates were highest using EZSPHERE plate type 4000-903, which has wells with a diameter of 800 μm and a depth of 400 μm. The proliferation rates were also the highest for this plate type. 3D cell cultures are used widely in research in various tissues, such as the vascular endothelium and cartilage, in an attempt to simulate the in vivo microenvironment. Better differentiation potential has been demonstrated in vitro compared with monolayer cultures of MSCs in spheroid cultures [26]. Previous work has shown that spheroid-forming NP cells have a greater ability to produce ECM [9,27] and are thus considered as promising cell types for regenerative and cellular therapeutics.

Our data support these previous findings, as EZSPHERE cultured cells exhibited a markedly higher *COL2A1* expression (on average 836-fold) compared with monolayer cultured NP cells. By contrast, *COL1A2* expression, as a negative NP marker [28], was markedly downregulated (on average 0.6-fold). The ECM produced by NP cells in human IVD tissue is classified mainly as type II collagen and proteoglycans, and IVD is not formed normally in type II collagen-deficient mice [29]. Therefore, type II collagen is thought to play an important role in the formation of the IVD, which suggests that NP cells that can be collected by this culture method should have excellent functionality. This supports our previously stated belief that the ability for colonization is an indicator of NP functionality [7].

We were particularly interested in the potential application of EZSPHERE plates as an alternative to the current standard of methylcellulose-based CFU-s cultures [13]. Culturing in EZSPHERE plates has an additional potential advantage because the diffusion gradient of the cultured cells and solution can be controlled more closely, which should improve the reproducibility and quality control for commercialization [30]. Conversely, methylcellulose-based culture leads to the formation of spheroid colonies of different sizes and, because of the high viscosity of the culture medium, cell recovery can be complicated. By contrast, the size of colonies obtained using EZSPHERE culture relies on the well's dimension, which produces relatively uniform colony sizes (Figure 1A). Moreover, the colonies are not maintained in a hydrogel, and the recovery of large quantities of colonies is much easier.

The EZSPHERE culture systems offer a promising culture device that could be scalable for production of CFU-s NP cells. As a goal for commercialization, for example, cells from one sample must be extracted and available for at least 100 people. This is important because IVD cell therapy products generally involve the transplantation of a total of 1–100 × 10^6 cells per disc (although an optimal concentration has not been determined) (15). For example, in our previous clinical trial [16], we obtained an average of 1.24 × 10^6 primary cells/7.92 g tissue per patient donor material (Table 3). Assuming that a cell therapy product requires 1 × 10^6 NP cells, ~100 times the cell proliferation effect is required. However, with the EZSPHERE method alone, the cell proliferation rate was 7.75-fold in our study, which would mean that only nine patients on average could be treated with cells from a single-batch production following EZSPHERE culture methods. Our data in the present study show that growth factor supplementation can markedly increase the NP cell proliferation rate. These findings are consistent with previous work showing that FGF2 has a Tie2 expression-enhancing effect on NP cells [14]. In our study, FGF2 supplementation had positive effect by increasing the cell proliferation rate by up to 62.5%. This would make it possible to manufacture NP cell products for about 77 patients from a single tissue sample. However, cells cultured with other growth factors showed a marked decrease expression of *COL2A1* and *ACAN*, which may be a disadvantage compared with the original colony assay culture. Further analysis and in vivo experimentation are needed to elucidate these effects.

Table 3. Details of previous Mochida et al. clinical trial of autologous NP cell transplantation. In the clinical trial of Mochida et al. [16], surgically removed NP tissue and obtained cells were cultured and administered to prevent adjacent degenerative disc disease. The average time required to recover the transplanted cells was 122.9 min using a similar procedure in this study. IVD: intervertebral disc, NPC: nucleus pulposus cell.

	Age	**Sex**	**IVD Tissue Amount (g)**	**Number of Collectied NPC**	**Number of Cultured NPC**	**Cultured Cell Recovery Time (min)**
1	21	M	8.2	1.25 × 10^6	7.2 × 10^6	115
2	28	M	4	0.93 × 10^6	5.85 × 10^6	113
3	28	M	10.9	3.42 × 10^6	10.80 × 10^6	105
4	27	M	8.9	1.20 × 10^6	5.8 × 10^6	174
5	29	M	3.6	0.71 × 10^6	3.36 × 10^6	100
6	28	M	7.7	1.0 × 10^6	3.02 × 10^6	120
7	21	M	11.9	1.05 × 10^6	2.88 × 10^6	120
8	29	M	6.6	0.5 × 10^6	2.48 × 10^6	125
9	21	F	9.1	0.93 × 10^6	2.63 × 10^6	115
10	25	M	8.3	1.43 × 10^6	2.80 × 10^6	142

The different types of cryopreservation solutions and exposure times used in the cryopreservation of cellular products are thought to affect the product quality [24]. Especially in undifferentiated cells such as MSCs, the maintenance and prevention of phenotype loss and changes in metabolism during storage are important. Reports have suggested inadequate therapeutic results can arise because of the effects of cryopreservation in clinical trials using MSCs [31,32]. Given our aim to deliver multipotent, undifferentiated NP progenitor cells with high Tie2 positivity, concerns were raised about the effects of cryopreservation on NP cells and their potency. In the trial reported by Mochida et al. [16], the collection of cultured cells before their administration in the operating room was performed in a similar fashion as in our experiments; that is, the NP cells were cryopreserved before transplantation.

The purpose of the current study was to identify the most appropriate cryopreservation solutions and freezing procedures by investigating their effects on NP cells over time. Of the 10 patients treated in the previous clinical trial [16], an average of 4.6×10^6 NP cells were administered following harvest from their respective coculture conditions. At the time of surgery, the cryostored NP cells were recovered, thawed, and transported from a cell-processing center to the operating room, and the waiting time before transplantation was on average 122.9 min, which meant that the cells were exposed to the cryopreserving agents for about 2 h. Our current results show that NP cells subjected to cryopreserving agents for up to 3 h had survival rates of ≤70% in SCB and SCBD-Free. Interestingly, cells kept on ice in CB1 retained a cell viability of 90% for up to 5 h.

Although CB1 has been shown to be optimal, for clinical translation, it is desirable to preserve and administer cells in solutions devoid of animal-derived serum or DMSO. Therefore, we examined six additional commercial cryopreserving preparations. We found that cell viability after thawing was maintained at ≥80% with CB1, CS10, and CP-1. CS10 was also able to maintain a high Tie2-expressing faction. In addition, CS10 is an animal component-free GMP-certified formula, making it potentially ssafer to administer directly to patients than CB1 when used as an OTS [23]. However, SCB is a product with an identical composition as CB1, but without animal components, and this resulted in a sharp decrease in viability at room temperature or on ice with time. Moreover, the CS10 formula does contain 10% DMSO, which showed a decrease in the number of cells after an incubation of ≥3 h incubation, which suggests that the collection and freezing of harvested cells should be performed in the shortest possible time. If we select one type of cell preservation solution for industrialization, CS10 is considered to be most optimal because it showed similar levels of cell viability maintenance as CB1 but does not contain any animal components.

The limitation of our study is that procurement of human-derived cell sources was limited, and some experiments had a limited sample size. Moreover, in this experiment, the cryopreservation period was only 24 h and long-term storage might have a more impactful effect on NP cells. Still, freezing and thawing process is likely the most important and damaging phase in cryopreservation, as such we believe this experiment is sufficient to compare the usefulness of solutions. Additionally, we did not further assess the impact of cryopreservatives on NP cell potency. Furthermore, a recent study by Croft et al. suggested that the type of cryopreservation medium, including CS10, had limited effects on NP cell multipotency [33]. The work by Hiraishi et al. suggested no effect on the in vivo regenerative potency of NP cells [24]. However, the effects on the potentially more sensitive Tie2 cells remains unclear [13]. Further advances in culture, storage, and logistics for NP cell-based cell therapeutics are needed to expand their potential as effective and marketable products in the clinic [17,25].

5. Conclusions

EZSPHERE culture plates enabled us to examine colony formation in cell culture without the use of methylcellulose or other hydrogel products. EZSPHERE culture methods produced high colony formation and increased proliferation rates, which could be augmented by FGF supplementation. Continued examination of the optimal culture methods is required for further development and commercialization. The derived cells should be assessed in vivo for their potential as a regenerative medicinal product. Our study showed that CS10 is presently, the best cryopreservation medium for retaining NP progenitor cell phenotype and viability when cells are kept on ice or even at room temperature for up to 3 h. Our findings provide information that may be useful for the translation of NP cell-based therapeutics to the clinic.

Author Contributions: Conceptualization, Y.N., D.S. and K.S.; methodology, K.S.; formal analysis, K.S., E.M., J.S. and T.W.; investigation, D.S.; resources, N.H., E.M. and T.W.; data curation, Y.N. and E.M.; writing—original draft preparation, K.S.; writing—review and editing, D.S., K.S., T.W., N.H., M.S., M.W. and J.S.; visualization, J.S.; supervision, M.S.; project administration, M.S. and M.W.; funding acquisition, D.S., M.S. and M.W. All authors have read and agreed to the published version of the manuscript.

Funding: This research received no external funding.

Institutional Review Board Statement: The study was conducted according to the guidelines of the Declaration of Helsinki, and approved by the Institutional Review Board for Clinical Research, Tokai University School of Medicine (protocol code 16R-051: 15 July 2016).

Informed Consent Statement: Informed consent was obtained from all subjects involved in the study.

Data Availability Statement: Data can be requested to the corresponding authors upon reasonable request.

Conflicts of Interest: The authors declare no conflict of interest.

References

1. Strine, T.W.; Hootman, J.M. US national prevalence and correlates of low back and neck pain among adults. *Arthritis Rheum.* **2007**, *57*, 656–665. [CrossRef]
2. Vos, T.; Allen, C.; Arora, M.; Barber, R.M.; Bhutta, Z.A.; Brown, A.; Carter, A.; Casey, D.C.; Charlson, F.J.; Chen, A.Z.; et al. Global, regional, and national incidence, prevalence, and years lived with disability for 310 diseases and injuries, 1990–2015: A systematic analysis for the Global Burden of Disease Study 2015. *Lancet* **2016**, *388*, 1545–1602. [CrossRef]
3. Oichi, T.; Taniguchi, Y.; Oshima, Y.; Tanaka, S.; Saito, T. Pathomechanism of intervertebral disc degeneration. *JOR Spine* **2020**, *3*, e1076. [CrossRef]
4. Frost, B.A.; Camarero-Espinosa, S.; Foster, E.J. Materials for the Spine: Anatomy, Problems, and Solutions. *Materials* **2019**, *12*, 253. [CrossRef]
5. O'Halloran, D.M.; Pandit, A.S. Tissue-engineering approach to regenerating the intervertebral disc. *Tissue Eng.* **2007**, *13*, 1927–1954. [CrossRef]
6. Antoniou, J.; Steffen, T.; Nelson, F.; Winterbottom, N.; Hollander, A.P.; Poole, R.A.; Aebi, M.; Alini, M. The human lumbar intervertebral disc: Evidence for changes in the biosynthesis and denaturation of the extracellular matrix with growth, maturation, ageing, and degeneration. *J. Clin. Investig.* **1996**, *98*, 996–1003. [CrossRef]
7. Weiler, C.; Nerlich, A.G.; Zipperer, J.; Bachmeier, B.E.; Boos, N. 2002 SSE Award Competition in Basic Science: Expression of major matrix metalloproteinases is associated with intervertebral disc degradation and resorption. *Eur. Spine J.* **2002**, *11*, 308–320. [CrossRef]
8. Le Maitre, C.L.; Freemont, A.J.; Hoyland, J.A. Localization of degradative enzymes and their inhibitors in the degenerate human intervertebral disc. *J. Pathol.* **2004**, *204*, 47–54. [CrossRef] [PubMed]
9. Sakai, D.; Nakamura, Y.; Nakai, T.; Mishima, T.; Kato, S.; Grad, S.; Alini, M.; Risbud, M.V.; Chan, D.; Cheah, K.S.; et al. Exhaustion of nucleus pulposus progenitor cells with ageing and degeneration of the intervertebral disc. *Nat. Commun.* **2012**, *3*, 1264. [CrossRef] [PubMed]
10. Guerrero, J.; Häckel, S.; Croft, A.S.; Albers, C.E.; Gantenbein, B. The effects of 3D culture on the expansion and maintenance of nucleus pulposus progenitor cell multipotency. *JOR Spine* **2020**, *4*, e1131.
11. Ishii, T.; Sakai, D.; Schol, J.; Nakai, T.; Suyama, K.; Watanabe, M. Sciatic nerve regeneration by transplantation of in vitro differentiated nucleus pulposus progenitor cells. *Regen. Med.* **2017**, *12*, 365–376. [CrossRef]
12. Li, X.C.; Tang, Y.; Wu, J.H.; Yang, P.S.; Wang, D.L.; Ruan, D.K. Characteristics and potentials of stem cells derived from human degenerated nucleus pulposus: Potential for regeneration of the intervertebral disc. *BMC Musculoskelet. Disord.* **2017**, *18*, 242. [CrossRef] [PubMed]
13. Sakai, D.; Schol, J.; Bach, F.C.; Tekari, A.; Sagawa, N.; Nakamura, Y.; Chan, S.C.W.; Nakai, T.; Creemers, L.B.; Frauchiger, D.A.; et al. Successful fishing for nucleus pulposus progenitor cells of the intervertebral disc across species. *JOR Spine* **2018**, *1*, e1018. [CrossRef] [PubMed]
14. Tekari, A.; Chan, S.C.W.; Sakai, D.; Grad, S.; Gantenbein, B. Angiopoietin-1 receptor Tie2 distinguishes multipotent differentiation capability in bovine coccygeal nucleus pulposus cells. *Stem Cell Res. Ther.* **2016**, *7*, 75. [CrossRef]

15. Schol, J.; Sakai, D. Cell therapy for intervertebral disc herniation and degenerative disc disease: Clinical trials. *Int. Orthop.* **2019**, *43*, 1011–1025. [CrossRef] [PubMed]
16. Mochida, J.; Sakai, D.; Nakamura, Y.; Watanabe, T.; Yamamoto, Y.; Kato, S. Intervertebral disc repair with activated nucleus pulposus cell transplantation: A three-year, prospective clinical study of its safety. *Eur. Cell Mater.* **2015**, *29*, 202–212, discussion 212. [CrossRef]
17. Sakai, D.; Schol, J. Cell therapy for intervertebral disc repair: Clinical perspective. *J. Orthop. Translat.* **2017**, *9*, 8–18. [CrossRef]
18. Gruber, H.E.; Ingram, J.A.; Norton, H.J.; Hanley, E.N., Jr. Senescence in cells of the aging and degenerating intervertebral disc: Immunolocalization of senescence-associated beta-galactosidase in human and sand rat discs. *Spine* **2007**, *32*, 321–327. [CrossRef]
19. Jiang, L.; Zhang, X.; Zheng, X.; Ru, A.; Ni, X.; Wu, Y.; Tian, N.; Huang, Y.; Xue, E.; Wang, X.; et al. Apoptosis, senescence, and autophagy in rat nucleus pulposus cells: Implications for diabetic intervertebral disc degeneration. *J. Orthop. Res.* **2013**, *31*, 692–702. [CrossRef]
20. Zhang, X.; Guerrero, J.; Croft, A.S.; Albers, C.E.; Häckel, S.; Gantenbein, B. Spheroid-Like Cultures for Expanding Angiopoietin Receptor-1 (aka. Tie2) Positive Cells from the Human Intervertebral Disc. *Int. J. Mol. Sci.* **2020**, *21*, 9423. [CrossRef]
21. Frauchiger, D.A.; Tekari, A.; May, R.D.; Dzafo, E.; Chan, S.C.W.; Stoyanov, J.; Bertolo, A.; Zhang, X.; Guerrero, J.; Sakai, D.; et al. Fluorescence-Activated Cell Sorting Is More Potent to Fish Intervertebral Disk Progenitor Cells Than Magnetic and Beads-Based Methods. *Tissue Eng. Part C Methods* **2019**, *25*, 571–580. [CrossRef]
22. Tanaka, M.; Sakai, D.; Hiyama, A.; Arai, F.; Nakajima, D.; Nukaga, T.; Nakai, T.; Mochida, J. Effect of cryopreservation on canine and human activated nucleus pulposus cells: A feasibility study for cell therapy of the intervertebral disc. *Biores. Open Access* **2013**, *2*, 273–282. [CrossRef]
23. Hiraishi, S.; Schol, J.; Sakai, D.; Nukaga, T.; Erickson, I.; Silverman, L.; Foley, K.; Watanabe, M. Discogenic cell transplantation directly from a cryopreserved state in an induced intervertebral disc degeneration canine model. *JOR Spine* **2018**, *1*, e1013. [CrossRef]
24. Morris, T.J.; Picken, A.; Sharp, D.M.C.; Slater, N.K.H.; Hewitt, C.J.; Coopman, K. The effect of Me(2)SO overexposure during cryopreservation on HOS TE85 and hMSC viability, growth and quality. *Cryobiology* **2016**, *73*, 367–375. [CrossRef]
25. Smith, L.J.; Silverman, L.; Sakai, D.; Le Maitre, C.L.; Mauck, R.L.; Malhotra, N.R.; Lotz, J.C.; Buckley, C.T. Advancing cell therapies for intervertebral disc regeneration from the lab to the clinic: Recommendations of the ORS spine section. *JOR Spine* **2018**, *1*, e1036. [CrossRef] [PubMed]
26. Yamaguchi, Y.; Ohno, J.; Sato, A.; Kido, H.; Fukushima, T. Mesenchymal stem cell spheroids exhibit enhanced in-vitro and in-vivo osteoregenerative potential. *BMC Biotechnol.* **2014**, *14*, 105. [CrossRef]
27. Lee, J.Y.; Hall, R.; Pelinkovic, D.; Cassinelli, E.; Usas, A.; Gilbertson, L.; Huard, J.; Kang, J. New use of a three-dimensional pellet culture system for human intervertebral disc cells: Initial characterization and potential use for tissue engineering. *Spine* **2001**, *26*, 2316–2322.
28. Risbud, M.V.; Schoepflin, Z.R.; Mwale, F.; Kandel, R.A.; Grad, S.; Iatridis, J.C.; Sakai, D.; Hoyland, J.A. Defining the phenotype of young healthy nucleus pulposus cells: Recommendations of the Spine Research Interest Group at the 2014 annual ORS meeting. *J. Orthop. Res.* **2015**, *33*, 283–293. [CrossRef] [PubMed]
29. Aszodi, A.; Chan, D.; Hunziker, E.; Bateman, J.F.; Fassler, R. Collagen II is essential for the removal of the notochord and the formation of intervertebral discs. *J. Cell Biol.* **1998**, *143*, 1399–1412. [CrossRef] [PubMed]
30. Silverman, L.I.; Flanagan, F.; Rodriguez-Granrose, D.; Simpson, K.; Saxon, L.H.; Foley, K.T. Identifying and Managing Sources of Variability in Cell Therapy Manufacturing and Clinical Trials. *Regen. Eng. Transl. Med.* **2019**, *5*, 354–361. [CrossRef]
31. Francois, M.; Copland, I.B.; Yuan, S.; Romieu-Mourez, R.; Waller, E.K.; Galipeau, J. Cryopreserved mesenchymal stromal cells display impaired immunosuppressive properties as a result of heat-shock response and impaired interferon-gamma licensing. *Cytotherapy* **2012**, *14*, 147–152. [CrossRef] [PubMed]
32. Moll, G.; Alm, J.J.; Davies, L.C.; von Bahr, L.; Heldring, N.; Stenbeck-Funke, L.; Hamad, O.A.; Hinsch, R.; Ignatowicz, L.; Locke, M.; et al. Do cryopreserved mesenchymal stromal cells display impaired immunomodulatory and therapeutic properties? *Stem Cells* **2014**, *32*, 2430–2442. [CrossRef] [PubMed]
33. Croft, A.S.; Guerrero, J.; Oswald, K.A.C.; Häckel, S.; Albers, C.E.; Gantenbein, B. Effect of different cryopreservation media on human nucleus pulposus cells' viability and trilineage potential. *JOR Spine* **2021**, *4*, e1140. [CrossRef] [PubMed]

MDPI
St. Alban-Anlage 66
4052 Basel
Switzerland
Tel. +41 61 683 77 34
Fax +41 61 302 89 18
www.mdpi.com

Applied Sciences Editorial Office
E-mail: applsci@mdpi.com
www.mdpi.com/journal/applsci

www.ingramcontent.com/pod-product-compliance
Lightning Source LLC
LaVergne TN
LVHW070645170726
843515LV00004B/328

* 9 7 8 3 0 3 6 5 2 7 5 4 3 *